树番茄

富硒措施研究

林立金　黄成毅　杨远祥◎主编

Study on Selenium Enrichment Measures in Tree Tomatoes

电子科技大学出版社
University of Electronic Science and Technology of China Press
·成都·

图书在版编目（CIP）数据

树番茄富硒措施研究 = Study on Selenium Enrichment Measures in Tree Tomatoes ：英文 / 林立金，黄成毅，杨远祥主编. -- 成都 ：成都电子科大出版社，2025. 7. -- ISBN 978-7-5770-1605-4

Ⅰ. S667.9

中国国家版本馆 CIP 数据核字第 20255QR483 号

树番茄富硒措施研究

Study on Selenium Enrichment Measures in Tree Tomatoes

林立金　黄成毅　杨远祥　主编

策划编辑　唐祖琴
责任编辑　赵倩莹
责任校对　饶定飞
责任印制　段晓静

出版发行　电子科技大学出版社
　　　　　成都市一环路东一段 159 号电子信息产业大厦九楼　邮编 610051
主　　页　www.uestcp.com.cn
服务电话　028-83203399
邮购电话　028-83201495

印　　刷　成都市火炬印务有限公司
成品尺寸　185 mm×260 mm
印　　张　13
字　　数　325 千字
版　　次　2025 年 7 月第 1 版
印　　次　2025 年 7 月第 1 次印刷
书　　号　ISBN 978-7-5770-1605-4
定　　价　55.00 元

编委会

主　编／林立金（四川农业大学）　黄成毅（四川农业大学）

杨远祥（四川农业大学）

副主编／金　鑫（四川农业大学）　廖人燕（雅安职业技术学院）

王　迅（四川农业大学）　杨茗琦（四川农业大学）

邹　瑶（四川农业大学）　朱嘉伟（四川农业大学）

参　编／王　进（四川农业大学）　李洪平（合江县公共检验检测中心）

肖云英（四川农业大学）　颜　寿（合江县公共检验检测中心）

张鸣飞（四川农业大学）　刘　继（成都市农林科学院）

孙国超（四川农业大学）　罗小莉（雅安职业技术学院）

徐湘婷（华中农业大学）　董依萍（华中农业大学）

余雪娜（四川农业大学）　郭奇亮（成都市农林科学院）

吕秀兰（四川农业大学）　蒋林宏（成都市农林科学院）

杨占彪（四川农业大学）　卢麒宇（成都市农林科学院）

李宇星（四川农业大学）　胡容平（四川省农业科学院植物保护研究所）

肖　巧（四川农业大学）　许　轲（四川省园艺作物技术推广总站）

石翔匀（四川农业大学）　周　琼（汉源县农业农村局果树站）

刘幸林（四川农业大学）　谢永东（成都市农林科学院）

唐旭民（四川农业大学）　张号楠（四川农业大学）

曹云凡（四川农业大学）　杨　倩（四川农业大学）

郭海霞（成都师范学院）　蒋　伟（成都师范学院）

Editorial Board

Xiulan Lv (Sichuan Agricultural University)

Linhong Jiang (Chengdu Academy of Agriculture and Forestry Sciences)

Zhanbiao Yang (Sichuan Agricultural University)

Qiyu Lu (Chengdu Academy of Agriculture and Forestry Sciences)

Yuxing Li (Sichuan Agricultural University)

Rongping Hu (Institute of Plant Protection, Sichuan Academy of Agricultural Sciences)

Qiao Xiao (Sichuan Agricultural University)

Ke Xu (Sichuan Horticultural Crops Technology Extension Station)

Xiangyun Shi (Sichuan Agricultural University)

Qiong Zhou (Fruit Tree Station of Hanyuan Agriculture and Rural Affairs Bureau)

Xinglin Liu (Sichuan Agricultural University)

Yongdong Xie (Chengdu Academy of Agriculture and Forestry Sciences)

Xumin Tang (Sichuan Agricultural University)

Haonan Zhang (Sichuan Agricultural University)

Yunfan Cao (Sichuan Agricultural University)

Qian Yang (Sichuan Agricultural University)

Haixia Guo (Chengdu Normal University)

Wei Jiang (Chengdu Normal University)

前　言

树番茄（*Cyphomandra betacea*）是一种集观赏与食用功能于一体的水果，原产于南美洲，在热带和亚热带地区均有引种，我国云南、贵州和西藏南部均有栽培。树番茄果实具有较高的营养价值，富含果胶、胡萝卜素、蛋白质、维生素、铁及其他多种矿质元素。因此，树番茄具有良好的发展前景。

随着人们生活水平的提高，人们在追求水果品种多样化的同时，也越来越注重其保健功能。硒是一种对人体极为重要的微量元素，具有抗癌作用，并且能增强人体对多种疾病（如心脑血管疾病、肝脏疾病和克山病）的抵抗力，同时还能延缓衰老、对有害化合物进行解毒以及增强免疫力。硒主要通过食物进入人体，而且食物中的硒只有一部分能被人体吸收。目前，提高农作物中硒含量的主要方法是向农作物施用不同形态的硒，但这种施硒方式可能带来硒污染的风险。我国大部分区域属低硒地质区，其农产品硒含量低，导致这些区域的居民面临硒摄入不足的风险。鉴于此，本课题组近年来针对如何提高树番茄硒富集的问题，研究了茶汤、植物生长调节剂、嫁接和混种等园艺措施对提高树番茄硒富集的效果，并对这些措施的作用机制进行了探讨。这些研究结果可为富硒树番茄的生产提供参考。

本书是将树番茄硒富集特性以及不同园艺措施对树番茄硒富集影响的相关成果汇编而成。全书共分为五章：第一章硒富集特性、第二章茶汤、第三章植物生长调节剂、第四章嫁接和第五章混种。编者在此特别感谢本书内容涉及的研究人员的辛勤付出，也对本书编写和出版过程中给予过关心和帮助的所有单位和个人表示感谢。

本书可作为从事浆果类果树科研和生产相关人员的参考用书。由于编者水平有限，书中难免有不妥之处，敬请各位读者和专家给予批评指正，并由衷感谢各位同人给予指点和帮助。

编　者

2025 年 6 月

Preface

Treetomato (*Cyphomandra betacea*) is a fruit that combines ornamental and edible functions. It is native to South America and has been introduced in tropical and subtropical regions, and has been cultivated in Yunnan, Guizhou, and southern Xizang of China. The fruits of tree tomato have high nutritional value, being rich in pectin, carotenoids, proteins, vitamins, iron, and many other mineral elements. Therefore, tree tomato has a promising development prospect.

With the improvement of people's living standards, while pursuing the diversification of fruit varieties, people are paying more and more attention to their health-care functions. Selenium is an extremely important trace element affecting the human body. It has anti-cancer effects and increases resistance to many diseases (e. g., cardiovascular and cerebrovascular diseases, liver disease, and Keshan disease), while also delaying aging, detoxifying harmful compounds, and enhancing immunity. Selenium enters human body mainly through food, and only a part of Se in food can be absorbed by the body. Currently, the main method to increase the selenium content in crops is to apply different forms of selenium to the crops. However, this method of selenium application may pose a risk of selenium pollution. Most areas in China are low-selenium geological areas, and the selenium content in agricultural products is low, resulting in the residents in these areas facing the risk of insufficient selenium intake. Therefore, to resolve the problem of how to increase the selenium enrichment of tree tomato, our research group studied the effects of horticultural measures such as tea infusion, plant growth regulators, grafting, and intercropping on improving the selenium enrichment of tree tomato in recent years, and explored the action mechanisms of these measures. These research results can provide a reference for the production of selenium-enriched tree tomato.

This book is compiled from the relevant achievements of the selenium enrichment characteristics oftree tomato and the influence of different horticultural measures on the selenium enrichment of tree tomato. The whole book is divided into five chapters: Chapter Ⅰ Selenium Accumulation Characteristics, Chapter Ⅱ Tea Infusion, Chapter Ⅲ Plant Growth Regulators, Chapter Ⅳ Grafting, and Chapter Ⅴ Intercropping. The editors would like to especially thank the researchers involved in the content of this book for their hard work, and also express the gratitude to all the institutions and individuals who have shown care and provided help during the compilation and publication of this book.

This book can serve as a reference for those engaged in scientific research and production related to berry fruit trees. Due to the limited time and the editors' limited capabilities, there may inevitably be some errors and omissions in the book. Editors sincerely hope that readers and experts will offer their criticisms and corrections, and editors are truly grateful to all colleagues for their guidance and assistance.

Editors

June, 2025

目　录

Chapter I
Selenium accumulation characteristics

1 Introduction

Selenium (Se) is an extremely important trace element affecting the human body (Xia, 2013). It has anti-cancer effects and helps increase resistance to many diseases (e. g., cardiovascular and cerebrovascular diseases, liver disease, and Keshan disease), while also delaying aging, detoxifying harmful compounds, and enhancing immunity (Xia, 2013; Zhao et al., 2007). Se enters human body mainly through food, and only a part of Se in food can be absorbed by the body. Different foods have different bioavailability of Se. For example, rice has 67%-76% Se bioavailability of total Se (Sun et al., 2017). Among commonly consumed foods, seafood and liver have the highest Se content, whereas grains have relatively low Se levels and fruits and vegetables have the lowest Se content (Iqbal et al., 2008; Wang, 2007). The long-term consumption of fish and meat with high Se content may lead to digestive problems and may be cost-prohibitive, and excessive Se supplementation will also lead to Se poisoning (Shi et al., 2005; Qu et al., 2010). Therefore, cheap, fresh, and delicious Se-containing vegetables and fruits from Se-rich areas and plant have become an important alternative source for this element (Yu et al., 2020; Huang et al., 2013; Alfthan et al., 2015). Although there are reports describing the Se content of vegetables and fruits, relatively few vegetables and fruits with high Se content have been identified (Wu et al., 2012; Temmerman et al., 2014; Zhang et al., 2009). Therefore, the continued screening for Se-rich vegetables and fruits is necessary.

Se is a non-essential element to plants, and on the basis of the considerable diversity in Se tolerance and absorption, plant species can be divided into the following three categories: primary Se accumulator (Se hyperaccumulator), secondary Se accumulator, and non-Se accumulator (Rosenfeld et al., 1964). The primary Se accumulators can grow in Se-rich soils and accumulate more than 1000 mg/kg Se. Examples of primary Se accumulators include *Astragalus* sp., *Stanleya* sp., and *Lecythis* sp. The secondary Se accumulators, which can accumulate Se at concentrations up to several hundred parts per million, include *Aster* sp., *Melilotus* sp., and *Brassica* sp. (Ellis et al., 2003; Shao et al., 2006; Jiang et al., 2007). Most vegetables and fruits are considered to be non-Se accumulators because they can accumulate relatively little Se and cannot grow under high-Se conditions (Zhang et al., 2012; Finley, 2005; Brown et al., 2010).

Cyphomandra betacea, which is a small Solanaceous tree or shrub, is a valuable ornamental plant species that produces sweet, sour, and juicy fruits rich in minerals, vitamins, and other trace elements (especially iron) (Yang, 2019; Guo et al., 2007). Although *C. betacea* originated in South America, it is now distributed in tropical and subtropical regions worldwide. In China, *C. betacea* has been cultivated in Yunnan province and southern Xizang as a source of fruits or vegetables (Zhang et al., 1998). In this study, we grew *C. betacea* seedlings in soils with varying Se concentrations and analysed the Se accumulation characteristics of the seedlings, and the tolerance of *C. betacea* seedlings to different Se concentrations (especially at high concentrations of Se) was also studied. The results of this investigation may be useful for developing new Se-rich *C. betacea* varieties.

2 Materials and methods

2.1 Materials

Mature seeds were collected from a *C. betacea* tree growing at the Chengdu Campus of Sichuan Agricultural University (30°71′ N, 103°87′ E) in October 2018. The seeds of *C. betacea* were sown in moist perlite and incubated in an artificial climate chamber under the following conditions: 14-h days (light intensity of 4,000 Lux) at 24 ℃ with 70% relative humidity and 10-h nights (0 Lux) at 20 ℃ with 90% relative humidity (Li et al., 2020). Approximately 1 month later, the seedlings of *C. betacea*, which were about 15 cm tall, were transplanted.

Fluvo-aquic soil collected from the Chengdu Campus of Sichuan Agricultural University in October 2018 was used in the experiments. The soil properties were the same as those reported by the previous studies (Xia et al., 2020; Li et al., 2019; Yao et al., 2019; Zhang et al., 2019). The total Se concentration of the soil was 0.35 mg/kg.

2.2 Experimental design

The experiment was conducted in a greenhouse at Chengdu Campus of Sichuan Agricultural University from October to December 2018. The soil samples were air-dried and passed through a 5-mm mesh in October 2018, after which 3.0 kg soil was added to polyethylene pots (15 cm tall, 18 cm diameter). A saturated Se solution (Na_2SeO_3) was added to the soil for final concentrations of 0, 5, 10, 25, or 50 mg/kg. The soil was mixed immediately and again after 4 weeks. Additionally, the soil moisture content was maintained at 80% of field capacity. When *C. betacea* seedlings were approximately 5 cm tall and had five expanded true leaves, they were transplanted to the potted soil. Four uniformly growing *C. betacea* seedlings were transplanted to each pot in November 2018. Each treatment was replicated five times in a completely randomized design. The soil moisture content was maintained at 80% of field capacity and the culture conditions were as follows: 14-h days (4,000 Lux) at 25 ℃ with 70% relative humidity and 10-h nights (0 Lux) at 20 ℃ with 90% relative humidity (Li et al., 2020). After a 1-month incubation, the fourth and fifth mature leaves of each seedling (counting from the top) were analysed regarding their photosynthetic pigment content. Specifically, the chlorophyll a, chlorophyll b, total chlorophyll, and carotenoid content were determined according to an ethanol/acetone extraction method involving the 663, 645, 652, and 470 nm wavelengths (Hao et al., 2004). Moreover, a total of 1.0 g aliquot of fresh leaf tissue was added to 6 mL extraction buffer (0.05 M potassium phosphate buffer containing 1 mM EDTA at pH 7.0) at 4 ℃, and the mixture was homogenized and then centrifuged at 11,000 × g for 20 min (Lin et al., 2020; Xu et al., 2020). The supernatant was used to determine the activities of the antioxidant enzymes superoxide dismutase (SOD), peroxidase (POD), and catalase (CAT) by the nitroblue tetrazolium photoreduction method, the guaiacol colorimetric method, and the potassium permanganate titration method, respectively (Hao et al., 2004). A total of 0.5 g aliquot of fresh leaf tissue was added to 6 mL distilled water, and the mixture was homogenized and then centrifuged at 4,000×g for 10 min. The supernatant was used to determine the soluble protein content according to a Coomassie brilliant blue staining method (Hao et al., 2004). Furthermore, DNA methylation was examined based on a previously described methylation-sensitive amplification polymorphism method and capillary gel electrophoresis (Xiong et al., 1999; Meng et al., 2010). A total of 24 primer pairs were screened, and 10 primer pairs (*E*32*HM*42, *E*32*HM*44, *E*32*HM*48, *E*37*HM*42, *E*37*HM*48, *E*38*HM*48, *E*40*HM*40, *E*40*HM*48, *E*59*HM*48, and *E*78*HM*42) with high proportions of amplified polymorphism were used for further amplifications (Table 1-1). The amplified products generated with the 10 primer pairs were divided into four categories (A, B, C, and D)and 15 subcategories. In Category A (Subcategories A1, A2, and A3), the DNA methylation patterns at CCGG sites were unchanged. In Category B (Subcategories B1, B2, B3, B4, and B5), the DNA methylation patterns at CCGG sites were changed by hypermethylation events. In Category C (Subcategories C1, C2, C3, C4, and C5), the DNA methylation patterns at the CCGG sites were changed by demethylation events. Finally, in Category D (Subcategories D1 and D2), the DNA methylation patterns were undetermined (Xiong et al., 1999; Meng et al., 2010).

Table 1-1 Adapters and primers

Adapter and primer	Sequence (5′-3′)
EcoR Ⅰ adaptor1	CTCGTAGACTGCGTACC
EcoR Ⅰ adaptor 2	AATTGGTACGCAGTCTAC
Hpa Ⅱ / *Msp* Ⅰ adaptor 1	GACGATGAGTCCTGAG
HpaI Ⅱ / *Msp* Ⅰ adaptor 2	CGCTCAGGACTCAT
EcoR Ⅰ pre-selective primer	GACTGCGTACCAATTC
Hpa Ⅱ / *Msp* I pre-selective primer	GATGAGTCCTGAGCGG
EcoR Ⅰ selective primer (*E*32)	GACTGCGTACCAATTCAAC
EcoR Ⅰ selective primer (*E*37)	GACTGCGTACCAATTCACG
EcoR Ⅰ selective primer (*E*38)	GACTGCGTACCAATTCACT
EcoR Ⅰ selective primer (*E*40)	GACTGCGTACCAATTCAGC
EcoR Ⅰ selective primer (*E*59)	GACTGCGTACCAATTCCTA
EcoR Ⅰ selective primer (*E*78)	GACTGCGTACCAATTCGTT
Hpa Ⅱ / *Msp* Ⅰ selective primer (*HM*40)	GATGAGTCCTGAGCGGAGC
Hpa Ⅱ / *Msp* Ⅰ selective primer (*HM*42)	GATGAGTCCTGAGCGGAGT
Hpa Ⅱ / *Msp* Ⅰ selective primer (*HM*44)	GATGAGTCCTGAGCGGATC
Hpa Ⅱ / *Msp* Ⅰ selective primer (*HM*48)	GATGAGTCCTGAGCGGCAC

Whole plants were harvested and divided into the roots, stems, and leaves, which were washed repeatedly with tap water and then rinsed three times with deionized water. After drying to a constant weight at 80 ℃, the dry weight of each plant part was determined. The dried plant parts were ground to fine powder for a chemical analysis of Se (volatile Se was not considered in this study). Samples (0.5 g) were digested in $HNO_3/HClO_4$ (4∶1, v/v), with HCl used as a reducing agent. The Se content in the roots, stems, and leaves were determined with the iCAP 6300 ICP-MS spectrometer (Thermo Scientific, Waltham, MA, USA) (Bao, 2000). The following formulae were used: Shoot biomass = stem biomass + leaf biomass; Shoot Se content = (Se content in stem × stem biomass + Se content in leaf × leaf biomass)/shoot biomass; Amount of Se extracted from plant = plant Se content × plant biomass (Zhang et al., 2010).

2.3 Statistical analyses

Data was subjected to a one-way analysis of variance with a least significant difference test (5% confidence level). All analyses were conducted with SPSS 20.0 (IBM, Chicago, IL, USA).

3 Results

3.1 Biomass of *C. betacea* seedlings

The root, stem, leaf, and shoot biomass were greater for *C. betacea* seedlings in soil with 5 mg/kg Se than for the control seedlings (Table 1-2). Specifically, the exposure to 5 mg/kg Se increased the root and shoot biomass by 48% and 6.4%, respectively, relative to the control biomass. In contrast, the 10, 25, and 50 mg/kg soil Se concentrations had no significant effect or decreased the root, stem, leaf, and shoot biomass of *C. betacea* seedlings. Thus, a low soil Se concentration (5 mg/kg) promoted *C. betacea* seedling growth, whereas high soil Se concentrations (greater than 5 mg/kg) had the opposite effect.

Table 1-2 Biomass of *C. betacea* seedlings

Se concentration/(mg/kg)	Root/(g/plant)	Stem/(g/plant)	Leaf/(g/plant)	Shoot/(g/plant)
0	0.242±0.004b	0.194±0.005b	0.696±0.008b	0.890±0.013b
5	0.357±0.004a	0.211±0.007a	0.736±0.023a	0.947±0.017a
10	0.148±0.003c	0.184±0.010b	0.551±0.015c	0.735±0.025c
25	0.122±0.006d	0.099±0.005c	0.383±0.012d	0.482±0.014d
50	0.120±0.004d	0.080±0.003d	0.242±0.005e	0.322±0.008e

Note: Different lowercase letters indicate significant differences among the treatments (least significant difference test, $P < 0.05$). Shoot biomass = stem biomass + leaf biomass.

3.2 Photosynthetic pigment content in *C. betacea* seedlings

Similar to the changes in biomass, the photosynthetic pigment (chlorophyll a, chlorophyll b, total chlorophyll, and carotenoid)content in *C. betacea* seedlings increased (relative to the control levels)only in response to the 5 mg/kg soil Se concentration (Table 1-3). At higher soil Se concentrations, the photosynthetic pigment content decreased or were not significantly affected.

Table 1-3 Photosynthetic pigment content in *C. betacea* seedlings

Se concentration/ (mg/kg)	Chlorophyll a/ (mg/g)	Chlorophyll b/ (mg/g)	Total chlorophyll/ (mg/g)	Carotenoid/ (mg/g)
0	2. 21±0. 04b	0. 74±0. 02b	2. 95±0. 03b	0. 44±0. 01b
5	2. 66±0. 05a	0. 89±0. 01a	3. 55±0. 05a	0. 52±0. 01a
10	2. 18±0. 05b	0. 71±0. 02c	2. 89±0. 04b	0. 43±0. 01b
25	2. 06±0. 04c	0. 68±0. 01d	2. 74±0. 03c	0. 41±0. 01c
50	1. 91±0. 03d	0. 65±0. 02e	2. 56±0. 02d	0. 38±0. 01d

Note: Different lowercase letters indicate significant differences among the treatments (least significant difference test, $P < 0.05$).

3.3 Antioxidant enzyme activities and soluble protein content of *C. betacea* seedlings

The 5 mg/kg soil Se concentration enhanced the SOD activity of *C. betacea* seedlings, whereas the 10, 25, and 50 mg/kg soil Se concentrations decreased the SOD activity relative to the control activity (Table 1-4). However, the increasing soil Se concentrations enhanced the POD and CAT activities of *C. betacea* seedlings, with the highest activities resulting from the 50 mg/kg Se concentration. The changes in the soluble protein content of *C. betacea* seedlings were similar to the changes in the SOD activity.

Table 1-4 Antioxidant enzyme activities and soluble protein content

Se concentration/ (mg/kg)	SOD activity/ (U/g)	POD activity/ [U/(g/min)]	CAT activity/ [U/(g/min)]	Soluble protein content/ (mg/g)
0	325±6. 15a	1,816±10. 81d	1. 86±0. 01e	11. 67±0. 25a
5	331±5. 33a	1,934±17. 50c	2. 13±0. 02d	11. 88±0. 20a
10	303±4. 22b	1,975±12. 82b	2. 18±0. 01c	9. 05±0. 09b
25	294±6. 41b	2009±19. 60b	2. 24±0. 02b	6. 73±0. 09c

continued table

Se concentration/ (mg/kg)	SOD activity/ (U/g)	POD activity/ [U/(g/min)]	CAT activity/ [U/(g/min)]	Soluble protein content/ (mg/g)
50	272±6. 52c	2310±31. 23a	2. 41±0. 03a	5. 22±0. 04d

Note: Different lowercase letters indicate significant differences among the treatments (least significant difference test, $P < 0.05$).

3.4 DNA methylation levels in *C. betacea* seedlings

The 5, 10, and 25 mg/kg soil Se concentrations increased the number of non-methylated sites and decreased the total number of methylated sites in *C. betacea* seedlings (relative to the control levels), whereas the 50 mg/kg soil Se concentration had the opposite effects (Table 1-5). The rank order of the soil Se concentrations for the number of non-methylated sites and the proportion of the total number of sites was 5 mg/kg > 10 mg/kg > 25 mg/kg > 0 mg/kg > 50 mg/kg. The rank order for the proportion of the total number of fully methylated sites was 25 mg/kg > 10 mg/kg ≈ 0 mg/kg ≈ 50 mg/kg > 5 mg/kg, whereas the rank order for the proportion of the total number of hemi-methylated sites was 50 mg/kg > 0 mg/kg > 25 mg/kg ≈ 5 mg/kg > 10 mg/kg. The rank order for the proportion of the total number of methylated sites was 50 mg/kg > 0 mg/kg > 25 mg/kg > 10 mg/kg > 5 mg/kg.

Table 1-5 DNA methylation levels

Se concentration/ (mg/kg)	Total site	Non-methylated site	Full-methylated site	Hemi-methylated site	Total-methylated site
0	637	102 (16%)	375 (59%)	160 (25%)	535 (84%)
5	665	157 (24%)	357 (53%)	151 (23%)	508 (76%)
10	649	121 (19%)	385 (59%)	143 (22%)	528 (81%)
25	650	112 (17%)	389 (60%)	149 (23%)	538 (83%)
50	643	93 (14%)	377 (59%)	173 (27%)	550 (86%)

Note: The number in bracket indicates the proportion of each site in total.

3.5 DNA methylation patterns in *C. betacea* seedlings

The various soil Se concentrations differentially affected the DNA methylation patterns of *C. betacea* seedlings (Table 1-6). The unchanged events (Category A) represented more than 45% of the product types that varied from the control. The rank order of the soil Se concentrations for Category B (hypermethylation events) was 10 mg/kg > 25 mg/kg > 5 mg/kg > 50 mg/kg (versus 0 mg/kg), whereas the rank order for Category C (demethylation events) was 10 mg/kg > 5 mg/kg > 25 mg/kg > 50 mg/kg (versus 0 mg/kg). Category D (unidentified events) represented less than 3% of the detected product types. There were fewer hypermethylated sites than demethylated sites for the 5 and 10 mg/kg soil Se concentrations (versus 0 mg/kg). Thus, the total methylated site ratios for 5 and 10 mg/kg were smaller than that for 0 mg/kg. There were a similar number of hypermethylated and demethylated sites for the 25 mg/kg soil Se concentration (versus 0 mg/kg). In contrast, there were more hypermethylated sites than demethylated sites for the 50 mg/kg soil Se concentration (versus 0 mg/kg).

Table 1-6 DNA methylation patterns

Band pattern	5, 10, 25 or 50		0		Number and frequency of pattern				
	Hpa Ⅱ	*Msp* Ⅰ	*Hpa* Ⅱ	*Msp* Ⅰ	5～0	10～0	25～0	50～0	
A					371 (50%)	340 (45%)	393 (54%)	434 (61%)	=
A1	1	1	1	1	165	150	168	183	
A2	1	0	1	0	83	59	77	101	
A3	0	1	0	1	123	131	148	150	
B					159 (21%)	183 (24%)	158 (21%)	133 (19%)	↑
B1	1	0	1	1	24	31	38	27	
B2	0	1	1	1	59	36	26	38	
B3	0	0	1	1	14	52	27	14	
B4	0	0	1	0	31	39	28	27	
B5	0	0	0	1	31	25	39	27	
C					194 (26%)	201 (27%)	160 (22%)	124 (17%)	↓
C1	1	1	1	0	34	34	36	25	
C2	1	1	0	1	56	63	43	25	
C3	1	1	0	0	34	15	15	14	
C4	1	0	0	0	39	52	33	32	
C5	0	1	0	0	31	37	33	28	
D					17 (3%)	29 (4%)	20 (3%)	20 (3%)	≠
D1	0	1	1	0	12	18	12	7	
D2	1	0	0	1	5	11	8	13	
Total					741	753	731	711	

Note: "1": band present, "0": band absent. " = " indicates that the patterns of DNA methylation at CCGG sites were unchanged; " ↓ " indicates that the patterns of DNA methylation at CCGG sites were changed by demethylation events in 5, 10, 25 or 50 mg/kg versus 0 mg/kg; " ↑ " indicates that the patterns of DNA methylation at CCGG sites were changed by hypermethylation events in 5, 10, 25 or 50 mg/kg versus 0 mg/kg; " ≠ " indicates the alternations of DNA methylation patterns that were unidentified by the methylation-sensitive amplification polymorphisms method. The number in bracket indicates the proportion of each pattern in total.

3.6 Se content in *C. betacea* seedlings

The rank order for Se content in the analysed plant tissues was roots > leaves > shoots > stems, except for the 0 and 5 mg/kg soil Se concentrations (Table 1-7). Increases in the soil Se concentration tended to increase the Se content in roots, stems, leaves, and shoots. At soil Se concentrations of 5, 10, 25, and 50 mg/kg, the Se content in roots were 30-, 77-, 145-, and 193-times that of the control level, respectively, whereas the Se content in shoots were 12-, 27-, 60-, and 84-times that of the control level, respectively.

Table 1-7 Se content in *C. betacea* seedlings

Se concentration/ (mg/kg)	Root/ (mg/kg)	Stem/ (mg/kg)	Leaf/ (mg/kg)	Shoot/ (mg/kg)
0	1. 18±0. 04e	0. 12±0. 01e	0. 53±0. 01e	0. 44±0. 01e
5	35. 83±0. 75d	5. 46±0. 09d	5. 23±0. 12d	5. 28±0. 11d
10	91. 34±1. 73c	10. 43±0. 29c	12. 42±0. 42c	11. 92±0. 39c
25	170. 56±4. 57b	24. 63±0. 38b	26. 74±0. 39b	26. 31±0. 40b
50	228. 01±5. 57a	29. 94±0. 68a	39. 38±0. 74a	37. 03±0. 70a

Note: Different lowercase letters indicate significant differences among the treatments (least significant difference test, $P < 0.05$). Se content in shoot = (Se content in stem × stem biomass + Se content in leaf × leaf biomass)/ shoot biomass.

3.7 Amount of Se extracted from *C. betacea* seedlings

Increases in the soil Se concentration were associated with an increasing trend in the amount of Se extracted from *C. betacea* seedling roots and stems (Table 1-8). The amount of Se extracted from leaves and shoots tended to increase at soil Se concentrations up to 25 mg/kg, but tended to decrease at higher soil Se concentrations. At soil Se concentrations of 5, 10, 25, and 50 mg/kg, the amount of Se extracted from roots increased by 44%, 47%,

72%, and 94%, respectively, relative to the control level, whereas the amount of Se extracted from shoots increased by 13%, 22%, 33%, and 31%, respectively, relative to the control level.

Table 1-8 Amount of Se extracted from *C. betacea* seedlings

Se concentration/ (mg/kg)	Root/ (μg/plant)	Stem/ (μg/plant)	Leaf/ (μg/plant)	Shoot/ (μg/plant)
0	0.29±0.01e	0.02±0.00d	0.37±0.01e	0.39±0.02e
5	12.79±0.39d	1.15±0.02c	3.85±0.12d	5.00±0.17d
10	13.52±0.46c	1.92±0.12b	6.85±0.39c	8.77±0.50c
25	20.83±1.52b	2.44±0.10a	10.24±0.42a	12.68±0.44a
50	27.35±0.72a	2.41±0.10a	9.52±0.14b	11.93±0.23b

Note: Different lowercase letters indicate significant differences among the treatments (least significant difference test, $P < 0.05$). Amount of Se extracted from plants = plant Se content × plant biomass.

4 Discussion

At an ideal concentration, Se can promote plant growth, but excessive amount of Se can be toxic to various plant species, including tomato, lettuce, spinach, cabbage, and carrot (Gregorio et al., 2006; Pezzarossa et al., 2007; Yu et al., 2007; Bicas et al., 1995). At a soil Se concentration of 20 mg/kg, the shoot and root biomass of mustard, lettuce, and spinach reportedly decrease (Duan et al., 2011). Additionally, common plants will be poisoned if the Se content in the plant body exceeds 50 mg/kg (Zayed et al., 1998). In the current study, only the 5 mg/kg soil Se concentration increased *C. betacea* seedling root and shoot biomass, whereas the other soil Se concentrations had the opposite effect. These results were similar to those of other studies (Wu et al., 2018; Li et al., 2018). Accordingly, a low soil Se concentration (5 mg/kg) may promote the growth of *C. betacea* seedlings, but higher concentrations (10 ~ 50 mg/kg) may adversely affect *C. betacea* seedling growth. In the future, the effects of more detailed low soil Se concentration range (0 ~ 5 mg/kg) on the growth of *C. betacea* seedlings should be studied.

Plants may contain selenoprotein structures similar to those of thioredoxin and iron thioproteins, which may regulate chlorophyll synthesis and photosynthesis (Liu et al., 2005; Ma et al., 2001). The effect of Se on the chlorophyll content might be related to the effects of Se on two important thiol-containing enzymes involved in chlorophyll biosynthesis (Padmaja et al., 1989). In the current study, the 5 mg/kg soil Se concentration increased the photosynthetic pigment (chlorophyll a, chlorophyll b, total chlorophyll, and carotenoid) content in *C. betacea* seedlings, whereas the other soil Se concentrations had no significant effect or decreased the abundance of these pigments. These observations were consistent with those of earlier studies (Wu et al., 2018; Li et al., 2018). The increased biosynthesis of photosynthetic pigments in *C. betacea* seedlings grown in soil containing 5 mg/kg Se may enhance photosynthetic activities and positively influence plant growth.

Se is similar to toxic metals, which can induce a strong redox reaction involving thiols. Divalent selenide is an oxidation catalyst that can oxidise thiols such as glutathione, reducing oxygen to generate free radical superoxide compounds (Spallholz et al., 2002; Spallholz, 1994), thereby affecting enzyme and protein activities. In plants, SOD, POD, and CAT form a reactive oxygen scavenging system targeting free radicals and peroxides, and their activities are related to the stress resistance of plants exposed to Se stress (Du et al., 2001; Du et al., 2007). In the current study, we observed that increases in the soil Se concentrations tended to decrease the SOD activity and soluble protein content of *C. betacea* seedlings, whereas the POD and CAT activities exhibited an increasing trend. These results were similar to those of a study of *Nasturtium officinale* treated with the same soil Se concentrations (Li et al., 2018).

Previous studies have confirmed that DNA methylation is an important epigenetic modification that helps control gene expression and maintain genome stability in higher plants (Zilberman, 2008; Steimer et al., 2000). Fluctuations in the internal and external environmental conditions cause DNA methylation patterns to change as part of an adaptive response (Ikeda et al., 2009; Fedoroff et al., 1995). Additionally, an exposure to stress often increases the DNA methylation levels in plants (Lin et al., 2016; Wu et al., 2013). However, Se stress (0.1 mm) can decrease the DNA methylation levels of tomato seedlings (Wang, 2014). In the present study, the 5, 10, and 25 mg/kg soil Se concentrations decreased the DNA methylation levels in *C. betacea* seedlings because of an increase in the demethylation patterns (versus 0 mg/kg), which is consistent with the findings of a previous study (Wang, 2014). The observed methylation changes may be related to the effects of Se-containing proteins on gene expression. However, the 50 mg/kg soil Se concentration increased the DNA methylation levels because of an increase in the hypermethylation patterns (versus 0 mg/kg), which differed from the observations of an earlier investigation (Wang, 2014), implying that the stress-induced changes to the DNA methylation levels (Lin et al., 2016; Wu et al., 2013) due to the high soil Se concentration exceeded the positive effects of Se. These results should be examined in greater detail in future studies.

The Se content in common plants is 0.02 ~ 1.5 mg/kg (Duan et al., 2011; Zayed et al., 1998). Additionally, the Se content in non-Se accumulators is generally less than 25 mg/kg (Shao et al., 2006). In this study, increasing soil Se concentrations caused Se content in *C. betacea* seedling to increase. Moreover, the amount of Se extracted from the roots also increased. These results were similar to those of previous studies (Wu et al., 2018; Li et al., 2018). The amount of Se extracted from the shoots tended to increase when the soil Se concentration did not exceed 25 mg/kg and decreased at soil Se concentrations greater than 25 mg/kg. Consistent findings were reported for an earlier study on *Solanum photeinocarpum* (Wu et al., 2018). In the future, the effects of more detailed low soil Se concentration range (0 ~ 5 mg/kg) on the Se accumulation characteristics of *C. betacea* seedlings should be studied. Compared with vegetable plants (Gregorio et al., 2006; Pezzarossa et al., 2007; Yu et al., 2007; Bicas et al., 1995), *C. betacea* seedlings can accumulate more Se, which may be relevant to future attempts at developing Se-rich *C. betacea* varieties.

References

[1] ALFTHAN G, EUROLA M, EKHOLM P, et al. Effects of nationwide addition of selenium to fertilizers on foods, and animal and human health in Finland: From deficiency to optimal selenium status of the population [J]. Journal of Trace Elements in Medicine and Biology, 2015, 31: 142-147.

[2] BAO S D. Soil agro-chemistrical analysis [M]. China Agriculture Press, Beijing, China, 2000.

[3] BICAS P A, DAOOD H G, Kodar I. Effect of Mo, Se, Zn, and Cr treatments on the yield, element concentration, and carotenoid content of carrot [J]. Journal of Agricultural and Food Chemistry, 1995, 43(3): 589-591.

[4] BROWN T A, SHRIFT A. Selenium: toxicity and tolerance in higher plants [J]. Biological Reviews, 2010, 57(1): 59-84.

[5] DU H L, FENG L R, NIU Z F, et al. The Effect of selenium on protection-enzyme activity and photosynthesis speed of lettuce [J]. Chinese Agricultural Science Bulletin, 2007, 23(5): 226-229.

[6] DU X M, YIN W X, ZHAO Y X, et al. The production and scavenging of reactive oxygen species in plants [J]. Chinese Journal of Biotechnology, 2001, 17(2): 121-125.

[7]DUAN M L, FU D D, WANG S S, et al. Effects of different selenite concentrations on plant growth, absorption and transportation of selenium in four different vegetables[J]. Acta Scientiae Circumstantiae, 2011, 31(3):658-665.

[8]ELLIS D R, SALT D E. Plants, selenium and human health[J]. Current Opinion in Plant Biology, 2003, 6(3):273-279.

[9]FEDOROFF N, SCHLÄPPI M, RAINA R. Epigenetic regulation of the maize Spm transposon[J]. Bioessays, 1995, 17:291-297.

[10]FINLEY J W. Selenium accumulation in plant foods[J]. Nutrition Reviews, 2005, 63(6):196-202.

[11]GREGORIO S D, LAMPIS S, MALORGIO F, et al. *Brassica juncea* can improve selenite and selenate abatement in selenium contaminated soils through the aid of its rhizospheric bacterial population[J]. Plant and Soil, 2006, 285(1-2):233-239.

[12] GUO B, ZHOU W, YE Q, et al. Biological characteristics and cultivation techniques of *Cyphomandra betacea* [J]. Guangdong Agricultural Sciences, 2007, (12):102-103.

[13]HAO Z B, CANG J, XU Z. Plant physiology experiment[M]. Harbin Institute of Technology Press, Harbin, China, 2004.

[14] HUANG Y, WANG Q, GAO J, et al. Daily dietary selenium intake in a high selenium area of Enshi, China[J]. Nutrients, 2013, 5(3):700-710.

[15] IKEDA Y, KINOSHITA T. DNA demethylation: a lesson from the garden[J]. Chromosoma, 2009, 118(1):37-41.

[16]IQBAL S, KAZI T G, BHANGER M I, et al. Determination of selenium content in selected Pakistani foods[J]. International Journal of Food Science and Technology, 2008, 43(2):339-345.

[17] JIANG Y B, JI H B, LI T T, et al. Advances of the research on the phytoremediation of the selenium polluted environment[J]. Bulletin of Mineralogy Petrology and Geochemistry, 2007, 26(1):98-104.

[18]LI H, WANG J, LIN L, et al. Effects of mutual grafting on cadmium accumulation characteristics of first post-generations of *Bidens pilosa* L. and *Galinsoga parviflora* Cav. [J]. Environmental Science and Pollution Research, 2019, 26:33228-33235.

[19]LI K Q, CHEN F B, LIN L J, et al. Effects of different concentrations of selenium on growth and selenium accumulation of *Nasturtium officinale* [J]. Journal of Sichuan Agricultural University, 2018, 36(4):488-493, 519.

[20] LI Z Y, ZHU J P, WANG Y X, et al. Effects of exogenous indole acetic acid on growth and cadmium accumulation of *Cyphomandra betacea* seedlings [J]. International Journal of Environmental Analytical Chemistry, 2020, 102(3):771-779.

[21] LIN L J, WU C F, WANG J, et al. Effects of reciprocal hybridization on cadmium accumulation in F1 hybrids of two *Solanum photeinocarpum* ecotypes [J]. Environmental Science and Pollution Research, 2020, 27(7):7120-7129.

[22] LIN L J, YANG D Y, WANG X, et al. Effects of grafting on the cadmium accumulation characteristics of the potential Cd-hyperaccumulator *Solanum photeinocarpum* [J]. Environmental Monitoring and Assessment, 2016, 188(2):1-11.

[23] LIU D H, ZHOU W B, ZHU D W, et al. General survey on physiological function of selenium in plant [J]. Journal of Mountain Agriculture and Biology, 2005, 24 (3): 253-259.

[24] MA Y H, DING R X, ZHANG J Z, et al. Interaction of selenium and sulfur in plants [J]. Plant Physiology Journal, 2001, 37(2):161-166.

[25] MENG H B, DU X, JIANG Y X, et al. Comparison between tetraploid turnip (*Brassica rapa*) and its diploid progenitor of DNA methylation under cadmium stress [J]. Journal of Nuclear Agricultural Sciences, 2010, 24(6):1297-1304.

[26] PADMAJA K, PRASAD D D K, PRASAD A R K. Effect of selenium on chlorophyll biosynthesis in mung bean seedlings [J]. Phytochemistry, 1989, 28 (12): 3321-3324.

[27] PEZZAROSSA B, PETRUZZELLI G, PETACCO F, et al. Absorption of selenium by *Lactuca sativa* as affected by carboxymethyl cellulose [J]. Chemosphere, 2007, 67 (2):322-329.

[28] QU L Z, YANG S J, GAO S, et al. Preliminary discussion on the effects of microelement selenium [J]. Chinese Agricultural Science Bulletin, 2010, 26(7):94-97.

[29] ROSENFELD I, BEATH OA. Selenium: geobotany, biochemistry, toxicity, and nutrition [M]. Academic Press, New York, USA, 1964.

[30] SHAO S X, ZHENG B S, WANG M S, et al. Review of selenium-hyperaccumulator plants [J]. Bulletin of Mineralogy, Petrology and Geochemistry, 2006, 25 (c1):31-33.

[31] SHI Y L, LUO S W. The advanced research of trace selenium [J]. Journal of Xiangfan Vocational and Technical College, 2005, 4(1):108-110.

[32] SPALLHOLZ J E, HOFFMAN D J. Selenium toxicity: cause and effects in aquatic birds[J]. Aquatic Toxicology, 2002, 57(1-2): 27-37.

[33] SPALLHOLZ J E. On the nature of selenium toxicity and carcinostatic activity[J]. Free Radical Biology and Medicine, 1994, 17(1): 45-64.

[34] STEIMER A, AMEDEO P, AFSAR K, et al. Endogenous targets of transcriptional gene silencing in *Arabidopsis*[J]. The Plant Cell, 2000, 12(7): 1165-1178.

[35] SUN G X, VAN de Wiele T, Alava P, et al. Bioaccessibility of selenium from cooked rice as determined in a simulator of the human intestinal tract (SHIME)[J]. Journal of the Science of Food and Agriculture, 2017, 97(11): 3540-3545.

[36] TEMMERMAN L D, WAEGENEERS N, THIRY C, et al. Selenium content of Belgian cultivated soils and its uptake by field crops and vegetables[J]. Science of the Total Environment, 2014, 468-469: 77-82.

[37] WANG B. The role and safety of selenium is not the past[J]. Health for the Elderly and Middle-Aged, 2007, 21(5): 48-49.

[38] WANG J W. The effect of selenium on DNA methylation of processing tomato under salt stress and its relationship with salt resistance [D]. Shihezi University, Xinjiang, China, 2014.

[39] WU C F, BAO R F, LIAO H P, et al. Selenium accumulation characteristics of wild vegetable *Solanum photeinocarpum*[J]. Journal of Sichuan Agricultural University, 2018, 36(6): 778-784.

[40] WU D, FANG K M, SHEN H F, et al. Comparative study on selenium-enriching ability of different types of vegetable in Leping City of Jiangxi [J]. Acta Agriculturae Jiangxi, 2012, 24(7): 23-24.

[41] WU R, WANG X, LIN Y, et al. Inter-species grafting caused extensive and heritable alterations of DNA methylation in Solanaceae plants [J]. PLOS ONE, 2013, 8(4): e61995.

[42] XIA H, WANG Y, LIAO M, et al. Effects of different rootstocks on cadmium accumulation characteristics of the post-grafting generations of *Galinsoga parviflora* [J]. International Journal of Phytoremediation, 2020, 22(1): 62-68.

[43] XIA Y M. Selenium[J]. Acta Nutrimenta Sinica, 2013, 35(3): 223-226.

[44] XIONG L Z, XU C G, MAROOF M A F, et al. Patterns of cytosine methylation in an elite rice hybrid and its parental lines, detected by a methylation-sensitive amplification polymorphism technique[J]. Molecular and General Genetics, 1999, 261(3): 439-446.

[45] XU D D, LI H Y, LIN L J, et al. Effects of carboxymethyl chitosan on the growth and nutrient uptake in *Prunus davidiana* seedlings[J]. Physiology and Molecular Biology of Plants, 2020, 26(4):661-668.

[46] YANG L C. Thinking on the present situation and development countermeasures of *Cyphomandra betacea* production in Tengchong City, Yunnan, China[J]. South China Agriculture, 2019, 13(5):113-114.

[47] YAO H, ZHANG F Q, QING M, et al. Effects of mutual grafting on the cadmium accumulation characteristics of two ecotypes of *Solanum photeinocarpum*[J]. International Journal of Phytoremediation, 2019, 21(5):503-508.

[48] YU T, HOU W L, HOU Q Y, et al. Safe utilization and zoning on natural selenium-rich land resources: a case study of the typical area in Enshi County, China[J]. Environmental Geochemistry and Health, 2020, 42(9):2803-2818.

[49] YU X Z, GU J D. Metabolic responses of weeping willows to selenate and selenite [J]. Environmental Science and Pollution Research, 2007, 14(7):510-517.

[50] ZAYED A, LYTLE C M, TERRY N. Accumulation and volatilization of different chemical species of selenium by plants[J]. Planta, 1998, 206(2):284-292.

[51] ZHANG D H, WANG Q P, MA X F. Prospective tropical fruit and vegetable plant *Cyphomandra betacea*[J]. Resource Development and Market, 1998, 14(5):209-210.

[52] ZHANG W, DI Z C. Research progress regarding the effect of exogenous selenium on vegetables[J]. Acta Horticulturae Sinica, 2012, 39(9):1749-1756.

[53] ZHANG X F, XIA H P, LI Z A, et al. Potential of four forage grasses in remediation of Cd and Zn contaminated soils[J]. Bioresource Technology, 2010, 101(6):2063-2066.

[54] ZHANG X Q, ZHANG H, WANG J, et al. Differences of accumulation and biological concentrator levels of selenium in different vegetables[J]. Jiangsu Agricultural Sciences, 2009, 37(4):170-171.

[55] ZHANG X Y, ZHANG F Q, WANG J, et al. Cutting after grafting affects the growth and cadmium accumulation of *Nasturtium officinale*[J]. Environmental Science and Pollution Research, 2019, 26(15):15436-15442.

[56] ZHAO H J, LIU P H, SUN Q Y, et al. Development of selenium on preventing and curing tumors[J]. Preventive Veterinary Medicine, 2007, 28(2):96-99.

[57] ZILBERMAN D. The evolving functions of DNA methylation[J]. Current Opinion in Plant Biology, 2008, 11(5):554-559.

Chapter Ⅱ
Tea infusion

1 Introduction

Selenium (Se) is one of the essential trace elements for human and animal life activities (Rayman, 2012). However, excessive Se production caused by the rock weathering, industrial activities, and agricultural activities has led to the serious Se-contamination in atmosphere, soil, and water (Alejandro et al., 2009; Paul et al., 2019). When the soil is contaminated by Se, the growth of plants is inhibited. It can even lead to plant death (Lanza et al., 2021; Ulhassan et al., 2019). Moreover, the excessive Se accumulation in plants can lead to oxidative and nitroxidative stress responses, which can also inhibit the growth of plants (Gupta et al., 2017; Jóźwiak et al., 2019). Thus, it is necessary to use appropriate methods to alleviate the effects of Se stress on plants.

Many agronomic measures, including the exogenous hormone application (Tang et al., 2020), grafting (Hu et al., 2021a; Huang et al., 2020), intercropping (Zhang et al., 2023), straw returning (Lin et al., 2017), and hybridization (Lin et al., 2020b), can promote or inhibit the uptake of mineral elements by plants. For the Se uptake, diethyl aminoethyl hexanoate can increase the Se content in grape seedlings (Liu et al., 2021a). Indole acetic acid can also increase the Se content in *Cyphomandra betacea* seedlings (Huan et al., 2021). In addition, the various rootstock combinations and grafting can affect the Se uptakes of plants differently (Huang et al., 2021; Huang et al., 2020). Various varieties of eggplant intercropping have different effects on the Se uptakes of eggplant (Pan et al., 2021). Applying the straw of *Pterocypsela laciniata* in the soil can promote the absorption and accumulation of Se in grape seedlings (Liu et al., 2021b). Hybridization of two different ecotypes of *Solanum photeinocarpum* can also promote the Se accumulation in F1 generation (Huang et al., 2021). Thus, diverse agronomic measures have different effects on the Se absorption and accumulation in plants.

Tea is one of the most popular non-alcoholic beverage. It is consumed by about half of the world's population (Szymczycha-Madeja et al., 2012). Tea is typically classified into white, black, green, dark, and yellow according to the fermentation degree (Hu et al., 2021b). Tea infusion is rich in the polyphenols, amino acids, caffeine, minerals, vitamins, and other physiological active substances, which are conducive to the human health and longevity (Syu et al., 2008). The polyphenols in tea have strong chelating properties and can form stable complexes with heavy metals to reduce the bioavailability of heavy metals (Huang, 2019; Duran et al., 2011; Amirhossein et al., 2016). The application of tea into soil decreased the cadmium (Cd) uptake and enhanced the resistance of pakchoi cabbage (Yang et al., 2018). However, the effect of tea infusion on the Se stress in plants has not been reported yet.

C. betacea belongs to Solanaceae family. Its fruit has a unique flavor. It is rich in a variety of minerals, pectin, and vitamin C. It has a high ornamental and edible value (Ramirez et al., 2019). Compared with the other Se-rich vegetable, the Se accumulation ability of *C. betacea* is still low (Lin et al., 2020a). Tea beverage processing produces a lot of tea residue, which can be used for further extraction of tea polyphenols or return to the fields (Liu, 2020). If application of tea infusion on *C. betacea*, the Se accumulation ability and commodity value of *C. betacea* may be improved. Therefore, in this study, four types of tea (black, green, white, and dark) infusions were foliar sprayed on the seedlings of *C. betacea* and the growth and Se uptake of *C. betacea* were then studied. The aim of this study was to find the appropriate tea infusion that not only could promote growth, but also could increase the Se content in *C. betacea*. The results of this study could provide a reference for the production of *C. betacea* in Se-contaminated soil.

2 Materials and Methods

2.1 Materials

Fluvo-aquic soil was collected from a farmland around the Chengdu Campus of Sichuan Agricultural University (30°42′N, 103°51′E). Its basic physicochemical properties have been described previously (Hu et al., 2021a).

The four types of tea were: black tea (product name: Second-level Gongfu Black Tea, produced by Leshan Yizhichun Tea Company), green tea (product name: Lvmaofeng, produced by Yaan Southern Yejia Tea Company), white tea (product name: Gongmei, produced by Fuding Guanzhongshan Tea Company), and dark tea (product name: Tibetan Tea, produced by Yaan Helong Tea Factory). Each tea (2 g) and distilled water (1,000 mL) were boiled on low heat for 30 min. After cooling, water was added to a final volume of 1, 200 mL, resulting in a tea:water ratio of 1:600 (w/v).

Seeds of *C. betacea* were harvested from the mature fruits of a 5-year-old *C. betacea* tree at the Chengdu Campus of Sichuan Agricultural University in September 2020, and stored in a 4 °C refrigerator before the experiment. In February 2021, seeds of *C. betacea* were sown in a moist substrate (perlite:vermiculite = 1:1) in an artificial climate chamber under the following conditions: 14-h days (light intensity of 10,000 Lux) at 24 °C with 70% relative humidity and 10-h nights (0 Lux) at 20 °C with 90% relative humidity (Li et al., 2022). Hoagland nutrient solution was used for irrigation every 3 days after emergence. When the seedlings of *C. betacea* were 15 cm tall on March 2021, they were transplanted into the pots.

2.2 Experimental design

The soil was dried and crushed, screened with a 3 mesh (6.72 mm), and put into a 20 cm × 21 cm (high × diameter) pot with 3.0 kg soil in each pot. Se was added to soil samples in the form of Na_2SeO_3 at a dose of 10 mg/kg (Hu et al., 2021a). Pots were placed in a transparent rain shelter for 1 month. Soil moisture content was kept at 80% of field capacity. Four uniform seedlings were transplanted into each pot in March 2021, and the seedlings were sprayed with distilled water (control) or tea (black, green, white or dark) infusions individually until the water droplets formed on leaves. Leaves were sprayed once every 7 days for four consecutive times until harvest. Each treatment was completed with three replicates (two pots for one replicate) with a total of 30 pots.

2.3 Sample analysis

Seven days after the last spraying (May, 2021), the fresh mature leaves (four to five from top to bottom) were harvested to measure the content of photosynthetic pigments (chlorophyll a, chlorophyll b, total chlorophyll, and carotenoid) and activities of antioxidant enzymes [superoxide dismutase (SOD), catalase (CAT), and peroxidase (POD)]. Photosynthetic pigment content was extracted with 4:1 (v/v) acetone-ethanol. Their absorbance levels were measured at various wavelengths (663, 645, and 470 nm) using a spectrophotometer (Summit, Shanghai, China). Activities of SOD, CAT, and POD were determined using the methods of nitrotetrazole chloride reduction, potassium permanganate titration, and guaiacol, respectively (Hao et al., 2004). Then, the whole plants were collected, washed with water three times, and divided into roots, stems, and leaves. After dried to constant weights at 75 ℃, the roots, stems, and leaves were weighed separately with an electronic balance. Dried samples were ground and sieved through a 0.15-mm mesh nylon sieve. Samples (1.000 g) were digested with 4∶1 (v/v) nitrate-perchloric acid until the solution was transparent (Lin et al., 2023). The solution was then added with deionized water to have a finale volume of 50 mL. After using filter paper for filtration, the concentrations of Se, potassium (K), and phosphorus (P) in various organs were determined with an iCAP6300 ICP spectrometer (Thermo Fisher Scientific, USA).

2.4 Statistical analyses

All statistical analyses in this study were conducted using SPSS 20.0 statistical software (SPSS Inc., Chicago, IL, USA). Data was normalized and subjected to a homogeneity test before one-way analysis of variance and Duncan's multiple range test ($P < 0.05$). Pearson correlation was used for correlation analysis, and the Pearson's correlation coefficient was visualized using Cytoscape (version 3.6.0) for Windows (The Cytoscape Consortium, New York, USA). Shoot biomass = stem biomass + leaf biomass; Se content in shoots = (Se content in stems × stem biomass + Se content in leaves × leaf biomass)/ shoot biomass; Se translocation factor (TF) = Se content in shoots/ Se content in roots (Rastmanesh et al., 2010).

3 Results

3.1 Biomass of *C. betacea*

Compared to the control, foliar spraying the white and dark tea infusions significantly increased the root, stem, leaf, and shoot biomass of *C. betacea* (Table 2-1). Although the black and green tea infusions also increased these values, such increases were not statistically significant. The highest biomass of root, stem, leaf, and shoot were produced after spraying the dark tea infusion, resulting in increases of 33.43% ($P < 0.05$), 35.92% ($P < 0.05$),

18.14% (P < 0. 05), and 24. 38% (P < 0. 05), respectively, compared with those of the control.

Table 2-1　Biomass of *C. betacea*

Treatment	Root/ (g/plant)	Stem/ (g/plant)	Leaf/ (g/plant)	Shoot/ (g/plant)
Control	1. 047±0. 021b	0. 952±0. 039c	1. 781±0. 087b	2. 732±0. 126c
Black tea	1. 060±0. 066b	0. 966±0. 023c	1. 803±0. 074b	2. 769±0. 097c
Green tea	1. 134±0. 073b	1. 007±0. 027c	1. 955±0. 083ab	2. 962±0. 110bc
White tea	1. 367±0. 082a	1. 157±0. 056b	2. 016±0. 096a	3. 173±0. 152ab
Dark tea	1. 397±0. 048a	1. 294±0. 040a	2. 104±0. 054a	3. 398±0. 014a

Note: Different lowercase letters indicate significant differences among the treatments (Duncan's multiple range test, P < 0. 05). Shoot biomass = stem biomass + leaf biomass.

3. 2　Photosynthetic pigment content in *C. betacea*

Compared to the control, spraying four kinds of tea infusions increased the chlorophyll a, chlorophyll b, total chlorophyll, and carotenoid content in C. betacea (Table 2-2). Spraying the dark tea infusion had the most significant effect in increasing the chlorophyll a, chlorophyll b, total chlorophyll, and carotenoid content [increases of 31. 84% (P < 0. 05), 47. 39% (P < 0. 05), 35. 10% (P < 0. 05), and 22. 01% (P < 0. 05), respectively, compared to the control]. However, spraying the tea infusion showed no significant effect on the chlorophyll a/b ratio of *C. betacea*, compared to the control.

Table 2-2　Photosynthetic pigment content

Treatment	Chlorophyll a/ (mg/g)	Chlorophyll b/ (mg/g)	Total chlorophyll/ (mg/g)	Carotenoid/ (mg/g)
Control	1. 564±0. 061c	0. 422±0. 020c	1. 986±0. 042c	0. 359±0. 017c
Black tea	1. 869±0. 054b	0. 553±0. 018b	2. 421±0. 071b	0. 401±0. 013b
Green tea	1. 963±0. 035ab	0. 581±0. 028ab	2. 545±0. 007ab	0. 440±0. 014a
White tea	2. 011±0. 015a	0. 591±0. 021ab	2. 602±0. 036a	0. 442±0. 012a
Dark tea	2. 062±0. 059a	0. 622±0. 017a	2. 683±0. 077a	0. 438±0. 018a

Note: Different lowercase letters indicate significant differences among the treatments (Duncan's multiple range test, P < 0. 05).

3. 3　Antioxidant enzyme activities of *C. betacea*

Spraying the white and dark tea infusions significantly enhanced SOD and CAT activities of *C. betacea* compared to the control (Table 2-3). Activities of SOD and CAT were the highest in the presence of dark tea infusion [increases of 19.11% ($P < 0.05$) and 21.66% ($P < 0.05$), respectively, in comparison with the control]. Compared to the control, the POD activities were increased 5.61% ($P < 0.05$), 11.53% ($P < 0.05$), and 14.35% ($P < 0.05$) by treatments with the black tea, white tea, and dark tea infusions, respectively. However, the green tea infusion decreased the POD activity.

Table 2-3 Antioxidant enzyme activities

Treatment	SOD activity/(U/g)	CAT activity/[mg/(g/min)]	POD activity/[U/(g/min)]
Control	169.26±9.74b	3.545±0.043c	3,583±37.02d
Black tea	171.72±8.12b	3.536±0.094c	3,784±18.18c
Green tea	181.57±6.49ab	3.712±0.048bc	3,488±29.41e
White tea	198.51±9.63a	3.862±0.114b	3,996±54.55b
Dark tea	201.61±2.28a	4.313±0.127a	4,097±16.97a

Note: Different lowercase letters indicate significant differences among the treatments (Duncan's multiple range test, $P < 0.05$).

3.4 P content in *C. betacea*

Spraying four kinds of tea infusions had no significant effects on the P content in roots or stems of *C. betacea* compared to the control (Table 2-4). For the P content in leaves, treatments were ranked as follows in the order of high P content to low P content: white tea infusion > green tea infusion > dark tea infusion > black tea infusion > control. Compared to the control, the P content in leaves treated with the white tea infusion was increased 13.44% ($P < 0.05$).

Table 2-4 P content

Treatment	Root/(g/kg)	Stem/(g/kg)	Leaf/(g/kg)
Control	1.662±0.078ab	0.959±0.040a	2.187±0.025c
Black tea	1.620±0.085b	0.993±0.019a	2.292±0.022bc
Green tea	1.790±0.052a	1.052±0.037a	2.412±0.088ab
White tea	1.741±0.054ab	1.038±0.045a	2.481±0.054a
Dark tea	1.772±0.013ab	1.025±0.042a	2.393±0.028ab

Note: Different lowercase letters indicate significant differences among the treatments (Duncan's multiple range test, $P < 0.05$).

3.5 K content in *C. betacea*

The K content in roots and leaves of *C. betacea* showed the following order: green tea infusion > white tea infusion > dark tea infusion > black tea infusion > control (Table 2-5). Compared to the control, all four tea infusions increased the K content in stems and leaves. In the presence of green tea infusion, the K content in stems and leaves was increased by 26.64% ($P < 0.05$), and 14.69% ($P < 0.05$), respectively, in comparison with those in the control.

Table 2-5 K content

Treatment	Root/ (g/kg)	Stem/ (g/kg)	Leaf/ (g/kg)
Control	6.825±0.283b	9.16±0.201d	13.55±0.281c
Black tea	6.900±0.212b	10.73±0.332bc	14.32±0.176b
Green tea	7.982±0.387a	11.60±0.194a	15.54±0.118a
White tea	7.931±0.196a	11.10±0.437ab	14.53±0.411b
Dark tea	7.722±0.094a	10.19±0.113c	14.46±0.310b

Note: Different lowercase letters indicate significant differences among the treatments (Duncan's multiple range test, $P < 0.05$).

3.6 Se content in *C. betacea*

Foliar spraying four types of tea infusion diminished the Se content in roots, stems, leaves, and shoots of *C. betacea* compared to the control (Table 2-6). The Se content in roots, stems, leaves, and shoots were decreased the most after foliar spraying the dark tea infusion [decreases of 33.11% ($P < 0.05$), 55.50% ($P < 0.05$), 26.61% ($P < 0.05$), and 40.07% ($P < 0.05$), respectively, compared with those of the control]. In addition, *C. betacea* treated with spraying the dark tea infusion diminished the TF in comparison with the control, whereas the other three treatments had no significant effects on this value.

Table 2-6 Se content and TF

Treatment	Root/ (mg/kg)	Stem/ (mg/kg)	Leaf/ (mg/kg)	Shoot/ (mg/kg)	TF/
Control	3.262±0.066a	0.373±0.018a	0.233±0.007a	0.282±0.002a	0.086±0.001a
Black tea	2.652±0.081b	0.260±0.013b	0.221±0.010ab	0.235±0.011b	0.089±0.007a

continued table

Treatment	Root/ (mg/kg)	Stem/ (mg/kg)	Leaf/ (mg/kg)	Shoot/ (mg/kg)	TF/
Green tea	2. 307±0. 067c	0. 203±0. 010c	0. 206±0. 006bc	0. 205±0. 007c	0. 089±0. 001a
White tea	2. 264±0. 057c	0. 172±0. 005d	0. 197±0. 005c	0. 188±0. 005d	0. 083±0. 000ab
Dark tea	2. 182±0. 043c	0. 166±0. 010d	0. 171±0. 009d	0. 169±0. 002e	0. 077±0. 002b

Note: Different lowercase letters indicate significant differences among the treatments (Duncan's multiple range test, $P < 0.05$). Se content in shoot = (Se content in stem × stem biomass + Se content in leaf × leaf biomass)/ shoot biomass. Translocation factor (TF) = Se content in shoot/ Se content in root.

3.7 Correlation analysis of Se, P, and K content

There were significant ($P < 0.01$) positive correlations among the Se content in roots, stems, and leaves (Figure 2-1). The Se content in roots and stems had a significant ($P < 0.01$) negative correlation with the K content in roots, stems, and leaves and the P content in stems and leaves. The Se content in leaves only had a significant ($P < 0.01$) negative correlation with the P content in leaves. The K content in leaves had a significant ($P < 0.01$) positive correlation with the P content in stems and leaves, and the K content in stems had a significant ($P < 0.01$) positive correlation with the P content in leaves.

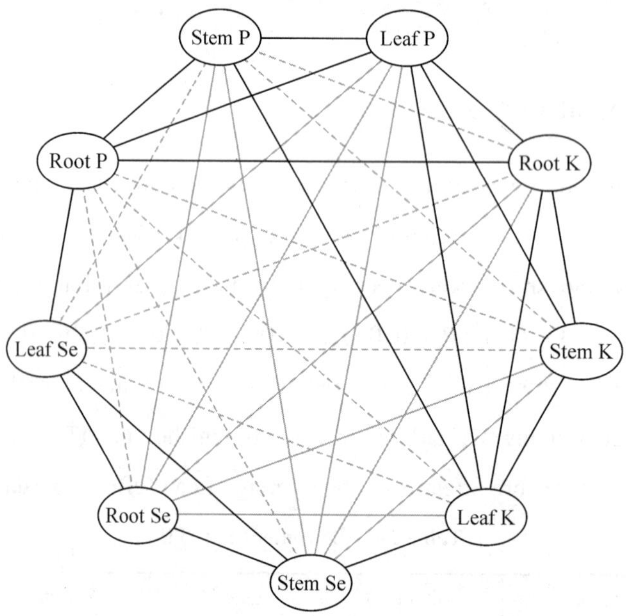

Note: Dashed lines represent no correlations ($P > 0.05$). Black and gray solid lines represent positive and negative correlations ($P < 0.01$), respectively.

Figure 2-1 Correlation analysis of Se with P and K content in roots, stems, and leaves

4 Discussion

Photosynthetic pigments play a significant role in photosynthesis. They are closely related to the biomass and yield of plant (Zepka et al., 2019). The high-level Se is toxic for photosynthetic pigments synthesis. It can decrease the biomass and yield of plants, including *C. betacea* (Lin et al., 2020a), wheat (Sharma et al., 2016), and rice (Wang et al., 2012). The main chemical components of tea infusion include the polyphenols, polysaccharides, proteins, amino acids, minerals, and so on (Szymczycha-Madeja et al., 2012; Karak et al., 2010). It has been reported that exogenous application of polysaccharide compound preparation, amino acid, and mineral nutrients can promote the growth of soybean (Bai et al., 2012), sweet pepper (Khan et al., 2018), and guava (Kumar et al., 2017), respectively. In the present study, the different tea infusions increased the chlorophyll a, chlorophyll b, total chlorophyll, and carotenoid content in *C. betacea*, but had no significant effect on the chlorophyll a/b ratio. The order of photosynthetic pigment content was: dark tea infusion > white tea infusion > green tea infusion > black tea infusion > control. These results suggested that the synergistic effect of various substances in tea infusion might have enhanced the photosynthetic capacity by increasing the photosynthetic pigment content, which are consistent with the previous studies of straw application (Lin et al., 2017; Liu et al., 2021b). In addition, compared with other treatments, foliar spraying the dark and white tea infusions were more conducive to the synthesis of photosynthetic pigments, indicating that the different tea infusions had various effects, which may be related to the content and compositions of tea (Syu et al., 2008). Moreover, the different tea infusions also increased the biomass of *C. betacea* to some extent, which is also consistent with the previous studies of straw application (Bai et al., 2012; Khan et al., 2018; Kumar et al., 2017). So, the tea infusion could promote the growth of *C. betacea* in Se-contaminated soil, which may be related to the polyphenols in tea (Syu et al., 2008; Huang, 2019). However, further studies are needed to verify this hypothesis.

Excessive amount of Se could induce the production of reactive oxygen species, which further stimulates the production of more antioxidants to resist the stress behavior of plants (Jiang et al., 2016), including *C. betacea* seedlings (Lin et al., 2020a) and *Pteris vittata* (Feng et al., 2012). Polyphenols are a biologically active group of tea components. They can up-regulate antioxidant enzymes indirectly (Bernatoniene et al., 2018). In addition, tea infusion is abundant in a variety of amino acids, which are the prerequisite of proteins, including proteins that are important for the synthesis, activity, gene expression, and Redox homeostasis of some enzymes (Batista-Silva et al., 2019). Previous studies have shown that foliar application of L-methionine and L-phenylalanine can effectively alleviate harmful effects of salinity on the growth and yield of tomato by enhancing the activities of antioxidant enzymes (Almas et al., 2021). In the present study, the different tea infusions showed various effects on the SOD, POD, and CAT activities of *C. betacea*. Notably, foliar spraying the dark and white tea infusions improved the antioxidant enzyme activity of *C. betacea* more significantly, which is consistent with the biomass and photosynthetic pigments of *C. betacea*. This suggests that only appropriate tea infusion could effectively enhance the resistance of *C. betacea*, thus effectively alleviate the adverse effects of Se stress on its growth. Types and concentrations of polyphenols and amino acids in various tea infusions are different (Hu et al., 2021b; Syu et al., 2008). This might explain why various treatments had different effects on the antioxidant system of *C. betacea*.

Application of Se can affect the nutrient element absorption of plants, including the P and K (Gupta et al., 2017; Jozwiak et al., 2019). In this study, all tea infusions except black tea enhanced the P uptake in leaves and K in roots, stems, and leaves of *C. betacea*. Tea infusion contains P, and S (Gezgin et al., 2006) and a number of mineral elements, including Al, Ca, K, Mg, Mn. The P and K as macronutrients can effectively activate the enzyme reaction, improve the ability to resist stress, and promote the growth of plants (Ma et al., 2008). Therefore, spraying tea infusion is equivalent to adding exogenous minerals, thus increasing P and K content in organs and improving the growth of *C. betacea*. In addition, this study showed that the Se content in roots, stems, leaves, and shoots of *C. betacea* had the following order: control > black tea infusion > green tea infusion > white tea infusion > dark tea infusion. Spraying the dark tea infusion decreased the TF of *C. betacea*. These results are consistent with the previous studies on the heavy metals (Huang, 2019; Duran et al., 2011; Amirhossein et al., 2016; Yang et al., 2018), indicating that the tea infusion stimulated the roots of *C. betacea* to secrete organic acids to chelate the soil Se and thereby decreasing the Se uptake (Fu et al., 2010). Moreover, correlation analysis showed that the content of P or K in each organ was mostly negatively correlated with the Se content. It has been found that the uptake of phosphate and selenite by crops may share the same phosphate transporters due to chemical similarity between phosphate and selenite (Li et al., 2008). Therefore, the phosphate can inhibit the selenite uptake by wheat (Wang et al., 2018). This was also verified in *C. betacea* in this study. Exogenous glutamic acid can significantly reduce content of Cd in the roots and stems of rice (jiang et al., 2020), demonstrating a positive role of amino acids in alleviating the stress. In view of this, we speculate that amino acids in tea infusion might be another key factor affecting Se absorption. In short, the decrease of Se content in *C. betacea* after treatment with tea infusion was due to a variety of factors. The mechanism merits further studies.

5 Conclusion

Among the four types of tea infusions, only foliar spraying the white and dark tea infusions significantly increased all of the biomass, photosynthetic pigment content, antioxidant enzyme activity of *C. betacea*, and thereby promoting the growth of *C. betacea*. The different tea infusions increased the K content in aboveground part and P in leaves of *C. betacea* to some extent. The different tea infusions decreased the Se content *C. betacea* to some extent. Correlation analysis showed that the content of P and K was negatively correlated with the Se content. So, foliar spraying tea infusion could reduce the Se uptake in *C. betacea*. The future studies should focus on investigating the mechanism of tea infusion reducing the Se uptake in *C. betacea*.

References

[1] ALEJANDRO F, LAURENT C. Selenium environmental cycling and bioavailability: a structural chemist point of view[J]. Reviews in Environmental Science and Biotechnology, 2009, 8(1):81-110.

[2] ALMAS H I, UN-NISA Z, ANWAR S, et al. Exogenous application of methionine and phenylalanine confers salinity tolerance in tomato by concerted regulation of metabolites and antioxidants[J]. Journal of Soil Science and Plant Nutrition, 2021, 21(4):3051-3064.

[3] AMIRHOSSEIN M, TAN S, SABA Y. Valorization of wasted black tea as a low-cost adsorbent for nickel and zinc removal from aqueous solution [J]. Journal of Chemistry, 2016, 2016(1):568-983.

[4] BAI W B, SONG J Q, GUO J Y, et al. Effects of plant polysaccharide compound agents on the photosynthetic characteristics and dry matter of soybean[J]. Chinese Journal of Applied Ecology, 2012, 23(7):1861-1868.

[5] BATISTA-SILVA W, HEINEMANN B, RUGEN N, et al. The role of amino acid metabolism during abiotic stress release [J]. Plant, Cell and Environment, 2019, 42 (5): 1630-1644.

[6] BERNATONIENE J, KOPUSTINSKIENE D M. The role of catechins in cellular responses to oxidative stress[J]. Molecules, 2018, 23(4): 965.

[7] DURAN C, OZDES D, GUNDOGDU A, et al. Tea-industry waste activated carbon, as a novel adsorbent, for separation, preconcentration and speciation of chromium [J]. Analytica Chimica Acta, 2011, 688(1): 75-83.

[8] FENG R W, WEI C Y. Antioxidative mechanisms on selenium accumulation in *Pteris vittata* L., a potential selenium phytoremediation plant [J]. Plant, Soil and Environment, 2012, 58(3): 105-110.

[9] FU X P, DOU C M, HU S P, et al. A review of progress in roles of organic acids on heavy metal resistance and detoxification in plants[J]. Chinese Journal of Plant Ecology, 2010, 34(11): 1354-1358.

[10] GEZGIN S, OZCAN M M, ATALAY E. Determination of minerals extracted from several commercial teas (*Camellia sinensis*) to hot water (infusion)[J]. Journal of Medicinal Food, 2006, 9(1): 123-127.

[11] GUPTA M, GUPTA S. An overview of selenium uptake, metabolism, and toxicity in plants[J]. Frontiers in Plant Science, 2017, 7(1): 2074.

[12] HAO Z B, CANG J, XU Z. Plant physiology experiment[M]. Harbin Institute of Technology Press, Harbin, China, 2004.

[13] HU R P, FAN Z H, CHEN Q H, et al. Effects of *Solanum nigrum* rootstock with different ploidies grafting on selenium accumulation of *Cyphomandra betacea* seedlings[J]. Chinese Agricultural Science Bulletin, 2021a, 37(2): 43-47.

[14] HU S, HE C, LI Y, et al. Changes of fungal community and non-volatile metabolites during pile-fermentation of dark green tea [J]. Food Research International, 2021b, 147(9): 110-472.

[15] HUAN Y M, YANG L, LIU Q, et al. Effects of indole acetic acid on the growth and selenium absorption characteristics of *Cyphomandra betacea* seedlings [J]. Acta Physiologiae Plantarum, 2021, 43(5): 1-8.

[16] HUANG K W, CAO Y L, ZHENG Y X, et al. Effects of reciprocal crosses on the growth and selenium accumulation in F1 generation of two ecotypes of *Solanum photeinocarpum*[J]. Journal of Yunnan Agricultural University (Natural Science), 2021, 36 (3): 377-383.

[17] HUANG K W, LI K Q, LIU J, et al. Effects of different rootstock combinations on physiological characteristics and selenium accumulation of *Nasturtium officinale* cuttings[J]. Acta Agriculturae Zhejiangensis, 2020, 32(3):447-454.

[18] HUANG K W, WANG Y, WEI X H, et al. Effects of mutual grafting *Solanum photeinocarpum* from two ecosystems on physiology and selenium absorption of their offspring under selenium stress[J]. Acta Physiologiae Plantarum, 2021, 43(7):96.

[19] HUANG Y. Research progress on removal of environmental pollutants with tea waste adsorption materials[J]. Journal of Hunan City University (Natural Science), 2019, 28(1):70-73.

[20] JIANG M, JIANG J, LI S, et al. Glutamate alleviates cadmium toxicity in rice via suppressing cadmium uptake and translocation[J]. Journal of Hazardous Materials, 2020, 384:121-319.

[21] JIANG Y, ZENG Z H, YANG Q S, et al. Selenium (Se)uptake and transformation mechanisms and physiological function in plant: A review[J]. Chinese Journal of Applied Ecology, 2016, 27(12):4067-4076.

[22] JOZWIAK W, POLITYCKA B. Effect of selenium on alleviating oxidative stress caused by a water deficit in cucumber roots[J]. Plants, 2019, 8(7):217.

[23] KARAK T, BHAGAT R M. Trace elements in tea leaves, made tea and tea infusion:A review[J]. Food Research International, 2010, 43(9):2234-2252.

[24] KHAN R I, HAFIZ I A, SHAFIQUE M, et al. Effect of pre-harvest foliar application of amino acids and seaweed (*Ascophylum nodosum*)extract on growth, yield, and storage life of different bell pepper (*Capsicum annuum* L.)cultivars grown under hydroponic conditions[J]. Journal of Plant Nutrition, 2018, 41(18):2309-2319.

[25] KUMAR J, KUMAR R, RAI R, et al. Influence of foliar application of mineral nutrients at different growth stages of guava[J]. Journal of Plant Nutrition, 2017, 40(5-8):656-661.

[26] LANZA M G D B, REIS D A R. Roles of selenium in mineral plant nutrition:ROS scavenging responses against abiotic stresses[J]. Plant Physiology and Biochemistry, 2021, 164(4):27-43.

[27] LI H F, MCGRATH S P, ZHAO F J. Selenium uptake, translocation and speciation in wheat supplied with selenate or selenite[J]. The New Phytologist, 2008, 178(1):92-102.

[28] LI Z, FAN R, PENG X, et al. Salicylic acid alleviates selenium stress and promotes selenium uptake of grpevine[J]. Physiology and Molecular Biology of Plants, 2022, 28:625-635.

[29] LIN L, LI Z, WANG J, et al. 24-epibrassinolide promotes selenium uptake in grapevine under selenium stress[J]. Scientia Horticulturae, 2023, 308:111564.

[30] LIN L J, LIAO M A, LV X L, et al. Addition of straw from hyperaccumulator plants to cadmium-contaminated soil increases cadmium uptake by loquat seedlings[J]. Environmental Monitoring and Assessment, 2017, 189(5):217.

[31] LIN L J, SUN J, CUI T H, et al. Selenium accumulation characteristics of *Cyphomandra betacea* (*Solanum betaceum*)seedlings[J]. Physiology and Molecular Biology of Plants, 2020a, 26(7):1375-1383.

[32] LIN L J, WU C F, WANG J, et al. Effects of reciprocal hybridization on cadmium accumulation in F1 hybrids of two *Solanum photeinocarpum* ecotypes[J]. Environmental Science and Pollution Research, 2020b, 27(7):7120-7129.

[33] LIU L, HAN J X, DENG L L, et al. Effects of diethyl aminoethyl hexanoate on the physiology and selenium absorption of grape seedlings[J]. Acta Physiologiae Plantarum, 2021a, 43(8):115.

[34] LIU P, HUANG K W, DAI J T, et al. Effects of applying *Pterocypsela laciniata* straw on growth and selenium accumulation of grape seedlings[J]. Journal of Hunan Agricultural University (Natural Sciences), 2021b, 47(1):35-39.

[35] LIU Z J. High-value comprehensive utilization of tea and tea residue[J]. Journal of Food Safety and Quality, 2020, 11(1):8-12.

[36] MA J M, CHENG L, CHANG J W, et al. Study on the effects of topdressing potassium sulfate on the quality and stress resistance of flu-cured tobacco in budding stage [J]. Journal of Anhui Agricultural Sciences, 2008(7):2792-2793.

[37] PAN S K, LU R H, LI H Y, et al. Mutual intercropping affects selenium uptake of eggplant seedlings[J]. International Journal of Environmental Analytical Chemistry, 2021, 101(15):2866-2875.

[38] PAUL T, SAHA N C. Environmental arsenic and selenium contamination and approaches towards its bioremediation through the exploration of microbial adaptations: A review[J]. Pedosphere, 2019, 29(5):554-568.

[39] RAMIREZ F, KALLARACKAL J. Tree tomato (*Solanum betaceum* Cav.) reproductive physiology: A review[J]. Scientia Horticulturae, 2019, 248(1):206-215.

[40] RASTMANESH F, MOORE F, KESHAVARZI B. Speciation and phytoavailability of heavy metals in contaminated soils in Sarcheshmeh area, Kerman Province, Iran[J]. Bulletin of Environmental Contamination and Toxicology, 2010, 85(5):515-519.

[41] RAYMAN M P. Selenium and human health[J]. Lancet, 2012, 379(9822):1256-1268.

[42] SHARMA S, GUPTA R, SINGH D. Variation in selenium tolerance, accumulation, and growth parameters of different wheat cultivars[J]. Communications in Soil Science and Plant Analysis, 2016, 47(2):203-212.

[43] SYU K Y, LIN C L, HUANG H C, et al. Determination of theanine, GABA, and other amino acids in green, oolong, black, and Pu-erh teas with dabsylation and high-performance liquid chromatography[J]. Journal of Agricultural and Food Chemistry, 2008, 56(17):7637-7643.

[44] SZYMCZYCHA-MADEJA A, WELNA M, POHL P. Elemental analysis of teas and their infusions by spectrometric methods[J]. Trends in Analytical Chemistry, 2012, 35(5):165-181.

[45] TANG Y, WANG L M, XIE Y D, et al. Effects of exogenous abscisic acid on the growth and cadmium accumulation of lettuce under cadmium-stress conditions[J]. International Journal of Environmental and Analytical Chemistry, 2020, 100(6):720-731.

[46] ULHASSAN Z, GILL R A, ALI S, et al. Dual behavior of selenium: Insights into physio-biochemical, anatomical and molecular analyses of four *Brassica napus* cultivars[J]. Chemosphere, 2019, 225(21):329-341.

[47] WANG P, HOU Z A, YE J, et al. Effects of different valence selenium on the absorption and transport of selenium in wheat[J]. Xinjiang Agricultural Sciences, 2018, 55(1):33-43.

[48] WANG Y D, WANG X, WONG Y S. Proteomics analysis reveals multiple regulatory mechanisms in response to selenium in rice[J]. Journal of Proteomics, 2012, 75(6):1849-1866.

[49] YANG Z, LIU L, LV Y, et al. Metal availability, soil nutrient, and enzyme activity in response to application of organic amendments in Cd-contaminated soil[J]. Environmental Science and Pollution Research, 2018, 25(3):2425-2435.

[50] ZEPKA L Q, JACOB-LOPES E, ROCA M. Catabolism and bioactive properties of chlorophylls[J]. Current Opinion in Food Science, 2019, 26(4):94-100.

[51] ZHANG R, HUANG Y, LIU Y, et al. Effects of mutual intercropping oncadmium accumulation of Solanum photeinocarpum Nakamura et Odashima and its post-grafting generations[J]. International Journal of Phytoremediation, 2023, 25(3):350-358.

Chapter Ⅲ
Plant growth regulators

Section Ⅰ Salicylic acid

1 Introduction

Selenium (Se)is an essential element for human body and deficiency in Se can lead to the development of various diseases(Reis et al., 2017). Humans are unable to synthesize Se and can only obtain it through food, with the amount of Se intake depending on the Se content present in crops (Winkel et al., 2015). The natural Se content in crops is typically low and may not meet the requirements for adequate Se intake in humans (Zhang et al., 2022). Therefore, the production of Se-enriched crops has become increasingly important (El-Ramady et al., 2016). Currently, the primary method for increasing Se content in crops involves applying different forms of Se to crops (Huan et al., 2021; Pan et al., 2021), with a few alternative methods available to enhance the Se-enriching capability of crops.

Salicylic acid (SA), a simple phenolic molecule, has been considered a plant hormone that regulates the development and defense of higher plants(Raskin, 1992). SA has various beneficial effects on plants, such as stimulating seed germination and increasing photosynthetic activity and growth rate (Cutt et al., 1992; Khan et al., 2003). SA can also act as an inducer to activate plant defense mechanisms by triggering the production of plant chemicals, leading to an increase in the total antioxidant polyphenol content of lettuce (Smoleń et al., 2019; Moreno-Escamilla et al., 2017; Saini et al., 2014). Additionally, SA can enhance plant resistance to both abiotic and biotic stress factors and improve mineral component absorption (Napoleão et al., 2017). The mechanism of SA in alleviating stress in plants primarily involves *P5CR* catalyzes the final reaction of proline biosynthesis, and the exogenous addition of SA can increase the expression of *P5CR* in leaves, ultimately enhancing the osmoregulatory capacity of plant cells and improving nutrient absorption from the soil (Feng et al., 2019; Li et al., 2004; Sun et al., 2022). Under drought stress, SA enhances the absorption of nitrogen, phosphorus, and potassium by plants, promoting overall plant growth and development while mitigating the damage caused by drought stress (Napoleão et al., 2017). Under cadmium (Cd) stress, SA promotes the growth of crops and also influences the accumulation of Cd in them (Tang et al., 2023; Liu et al., 2021). In the case of hyperaccumulator or accumulator plants, SA promotes the accumulation of Cd in these plants (Zheng et al., 2022). Under Se stress, SA can also increase Se content and promote the growth of grapevines (Li et al., 2022). However, the effects of SA on Se uptake in rice differ, with SA reducing Se uptake and alleviating Se-phytotoxicity in rice to enhance growth (Mostofa et al., 2020). Moreover, SA has varying effects on the Se content in different organs of potatoes (Smoleń et al., 2018). Therefore, the effects of SA on the Se absorption of different crops can vary.

Cyphomandra betacea, a tree with sweet, sour, and juicy fruit, is rich in vitamins, minerals, and other beneficial trace elements, and has high ornamental and edible value (Yang, 2019). However, the Se accumulation capability of *C. betacea* is low (Lin et al., 2020). Applying SA to *C. betacea* may improve its Se content and commercial value. Therefore, the effects of different concentrations of SA on the growth and Se accumulation of *C. betacea* seedlings was studied under Se stress. The aim of this study was to screen out the optimal SA concentration for increasing the Se content in *C. betacea* seedlings, and provide a reference for producing the Se-enriched fruits of *C. betacea*.

2　Materials and methods

2.1　Materials

Mature seeds of *C. betacea* were collected from a 5-year-old *C. betacea* tree at the Chengdu Campus of Sichuan Agricultural University (30° 71′ N, 103° 87′ E) in September 2019. The seeds were sown in a tray containing moist perlite and placed in an artificial climate chamber. The Hoagland nutrient solution was watered every three days to aid in the growth of seedlings. The artificial climate chamber was regulated to provide optimal conditions for growth, including 14-hour days at 25 ℃ with a relative humidity of 70% and 4000 Lux of light, followed by 10-hour nights at 20 ℃ with a relative humidity of 90% and no light (Li et al., 2022). Once the seedlings of *C. betacea* reached a height of 15 cm, they were transplanted to trays for continued growth and treatment.

2.2 Experimental design

The seedlings of *C. betacea* with strong and consistent growth were selected and transferred to nine-hole trays (14.5 cm in length, 14.5 cm in width, and 10 cm in height) containing perlite in October 2019. A total of 6 *C. betacea* seedlings were transplanted into each tray and kept in an artificial climate chamber with the same as the seedling nurturing conditions. Subsequently, different concentrations of SA [0 (control), 100, 150, 200, and 250 mg/L] were sprayed on both sides of the leaves until the solution began to drip, respectively (Li et al., 2022). Each tray received 25 mL of SA solution and was sprayed once every 7 days, for a total of 4 sprays. Each treatment was replicated three times, with one tray as a replicate. Concurrently, a concentration of 0.1 mg/L Se in the form of sodium selenite was added directly to the Hoagland nutrient solution every three days (Li et al., 2022). Throughout the seedling growth process, the position of the trays was irregularly changed to minimize the effects of the marginal conditions.

2.3 Determination of indicators

The net photosynthetic rate (Pn), transpiration rate (Tr), stomatal conductance (Gs), and intercellular CO_2 concentration (Ci) of the sixth mature leaf (counting from the top) of each plant were measured using the Li-6400 portable photosynthesis analyzer (LI-COR Inc., Lincoln, NE, USA). The fifth mature leaf of each plant was used to determine the activities of antioxidant enzymes, i. e., superoxide dismutase (SOD), peroxidase (POD), and catalase (CAT) through the nitroblue tetrazolium photoreduction, guaiacol colorimetric, and potassium permanganate titration methods, respectively, following the protocols established by Hao et al. (Hao et al., 2024). The soluble protein content was determined by the Coomassie brilliant blue staining method (Hao et al., 2024). The sixth mature leaf of each plant was selected to determine the content of photosynthetic pigments (chlorophyll a, chlorophyll b, carotenoid, and total chlorophyll) through the ethanol and acetone extraction method (Hao et al., 2024). Subsequently, the whole plant was harvested, divided into roots and shoots, and washed three times with tap water and deionized water. The plants were dried to a constant weight at 80 ℃ after being sterilized at 110 ℃ for 15 minutes. Then, the dry weight (biomass) of the different parts was measured. The dried plant parts were ground into a fine powder for Se chemical analysis. A 0.5 g sample was digested in $HNO_3/HClO_4$ (4:1, v/v), and HCl was used as a reducing agent. The Se content in roots and shoots were determined using the iCAP 6300

ICP-MS spectrometer (Thermo Scientific, Waltham, MA, USA)(Bao, 2000).

2.4 Statistical analyses

The data were analyzed using one-way analysis of variance and the Duncan's multiple range test at a 5% confidence level. Pearson's correlation was employed to examine the correlations among all parameters, while grey relational analysis was used to analyze the grey correlations of the different parameters with the shoot Se content (Lin et al., 2023). All analyses were performed using SPSS 20.0(IBM, Inc., Armonk, NY, USA) software.

3 Results

3.1 Biomass of *C. betacea* seedlings

With the increased SA concentration, the biomass of root and shoot of *C. betacea* seedlings increased under Se stress (Figure 3-1 and Figure 3-2). The biomass of each organ showed an increasing trend when the SA concentration was lower than 150 mg/L and a decreasing trend when the SA concentration was higher than 150 mg/L. Compared to the control, the concentrations of 100, 150, 200, and 250 mg/L of SA increased the root biomass by 35.16%, 52.55%, 45.70%, and 30.28%, respectively, and increased the shoot biomass by 6.82%, 37.68%, 27.75%, and 15.54%, respectively.

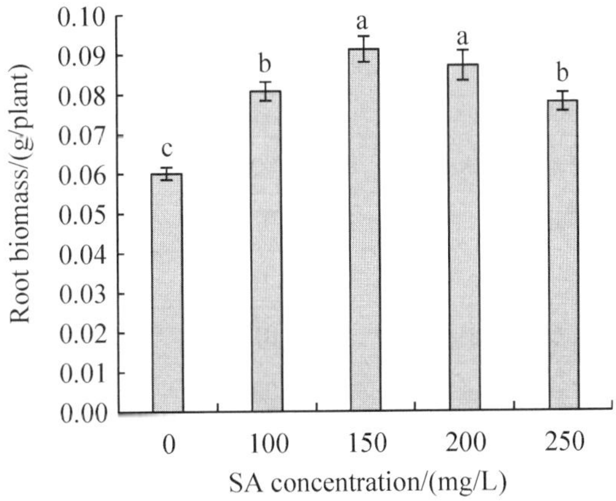

Note: Different lowercase letters above a column indicate significant differences based on one-way analysis of variance in SPSS 20.0 followed by the Duncan's multiple range test at the 5% confidence level.

Figure 3-1 Root biomass of *C. betacea* seedlings

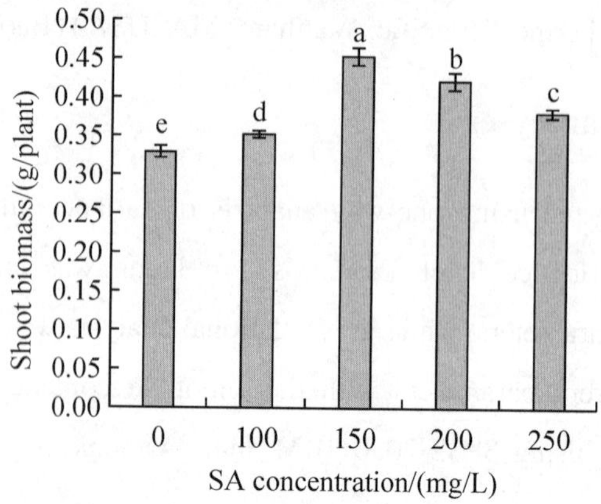

Note: Different lowercase letters above a column indicate significant differences based on one-way analysis of variance in SPSS 20.0 followed by the Duncan's multiple range test at the 5% confidence level.

Figure 3-2 Shoot biomass of *C. betacea* seedlings

3.2 Photosynthetic pigment content of *C. betacea* seedlings

All SA treatments led to an increase in the levels of chlorophyll a, chlorophyll b, carotenoid, and total chlorophyll in *C. betacea* seedlings exposed to Se stress (Table 3-1). Specifically, when the concentration of SA was 150 mg/L, the levels of chlorophyll a, chlorophyll b, carotenoid, and total chlorophyll reached their peak value, showing an increase of 70.16%, 92.71%, 146.76%, and 74.68%, respectively, compared to the control.

Table 3-1 Photosynthetic pigment content in *C. betacea* seedlings

SA concentration/(mg/L)	Chlorophyll a/(mg/g)	Chlorophyll b/(mg/g)	Carotenoid/(mg/g)	Total chlorophyll/(mg/g)
0	0.744±0.035d	0.192±0.009d	0.139±0.005e	0.936±0.026d
100	0.935±0.028c	0.277±0.014c	0.198±0.009d	1.212±0.042c
150	1.266±0.034a	0.370±0.017a	0.343±0.011a	1.635±0.049a
200	1.250±0.011a	0.340±0.011b	0.278±0.007b	1.590±0.010a
250	1.151±0.052b	0.321±0.008b	0.225±0.006c	1.473±0.059b

Note: Different lowercase letters within a column indicate significant differences based on one-way analysis of variance in SPSS 20.0 followed by the Duncan's multiple range test at the 5% confidence level.

3.3 Photosynthetic characteristics of *C. betacea* seedlings

Under Se stress, all of the SA treatments resulted in increases in the Pn, Gs, Ci, and Tr of *C. betacea* seedlings to some degree (Table 3-2). When the SA concentration was at 150 mg/L, both Pn and Tr reached their peak value, showing an increase of 167.23% and

174.93% respectively compared to the control. At an SA concentration of 200 mg/L, Gs reached its maximum value, showing a 198. 79% increase compared to the control. Only the 150 mg/L SA treatment resulted in a higher Ci compared to the control, showing a 13.18% increase.

Table 3-2 Photosynthetic characteristics of *C. betacea* seedlings

SA concentration/ (mg/L)	Pn/ [μmol CO_2/(m^2/s)]	Gs/ [mol H_2O/(m^2/s)]	Ci/ (μmol CO_2/mol)	Tr/ [mmol H_2O/(m^2/s)]
0	2. 014±0. 041e	0. 0165±0. 0002d	188. 2±3. 63b	0. 375±0. 006e
100	2. 983±0. 033d	0. 0253±0. 0006c	194. 5±5. 18b	0. 520±0. 011d
150	5. 382±0. 053a	0. 0477±0. 0011a	213. 0±9. 62a	1. 031±0. 015a
200	5. 083±0. 064b	0. 0493±0. 0014a	205. 8±8. 76ab	1. 006±0. 009b
250	4. 345±0. 083c	0. 0385±0. 0012b	197. 3±5. 69b	0. 836±0. 017c

Note: Different lowercase letters within a column indicate significant differences based on one-way analysis of variance in SPSS 20. 0 followed by the Duncan's multiple range test at the 5% confidence level. Pn = net photosynthetic rate, Tr = transpiration rate, Gs = stomatal conductance, Ci = intercellular CO_2 concentration.

3. 4 Antioxidant enzyme activity and soluble protein content in *C. betacea* seedlings

The activities of SOD, CAT, and POD in *C. betacea* seedlings were increased by SA treatments under Se stress, as shown in Table 3-3. The activities of SOD, CAT, and POD exhibited an increasing trend when the SA concentration was lower than 150 mg/L and a decreasing trend when the SA concentration exceeded 150 mg/L. At the SA concentration of 150 mg/L, the activities of SOD, CAT, and POD reached their maximum value, showing increases of 30. 52%, 18. 56%, and 49. 39%, respectively, compared to the control. In terms of soluble protein content, the concentrations of 100, 150, 200, and 250 mg/L SA increased it by 8. 80%, 23. 90%, 12. 66%, and 9. 62%, respectively, compared to the control.

Table 3-3 Antioxidant enzyme activity and soluble protein content in *C. betacea* seedlings

SA concentration/ (mg/L)	SOD activity/ (U/g)	CAT activity/ [mg/(g/min)]	POD activity/ [U/(g/min)]	Soluble protein content/ (mg/g)
0	205. 1±8. 74d	5. 360±0. 052d	988±19. 29e	79. 21±1. 40c
100	226. 6±10. 42c	5. 413±0. 068cd	1, 058±23. 29d	86. 18±1. 70b
150	267. 7±11. 57a	6. 355±0. 049a	1, 476±25. 45a	98. 14±2. 74a
200	255. 1±10. 36ab	5. 815±0. 081b	1, 375±30. 65b	89. 24±2. 40b

continued table

SA concentration/(mg/L)	SOD activity/(U/g)	CAT activity/[mg/(g/min)]	POD activity/[U/(g/min)]	Soluble protein content/(mg/g)
250	244.7±5.18b	5.531±0.042c	1,216±24.48c	86.83±3.07b

Note: Different lowercase letters within a column indicate significant differences based on one-way analysis of variance in SPSS 20.0 followed by the Duncan's multiple range test at the 5% confidence level.

3.5 Se content in various organs of *C. betacea* seedlings

The content of root Se and shoot Se in *C. betacea* seedlings was increased by the different SA concentrations under Se stress (Figure 3-3 and Figure 3-4). The Se content in various organs increased when the SA concentration was lower than 150 mg/L, but decreased when the concentration exceeded 150 mg/L. The root Se content showed a linear relationship with the SA concentration, while the shoot Se content exhibited a quadratic polynomial relationship with the SA concentration. Compared to the control, the concentrations of 100, 150, 200, and 250 mg/L SA increased the root Se content by 21.65%, 29.74%, 29.02%, and 23.78%, respectively, and increased the shoot Se content by 12.23%, 30.69%, 25.42%, and 14.98%, respectively.

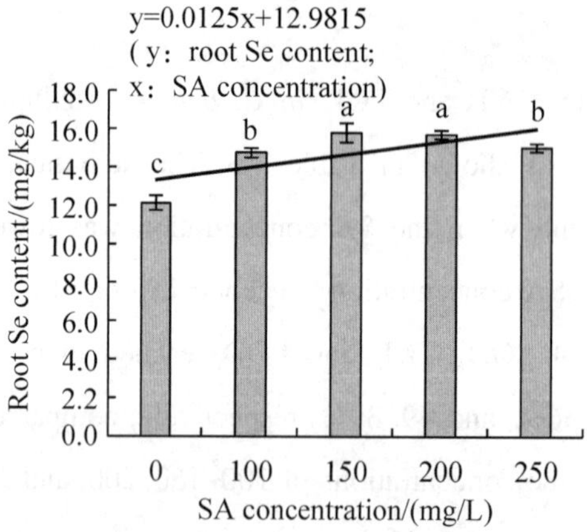

Note: Different lowercase letters above a column indicate significant differences based on one-way analysis of variance in SPSS 20.0 followed by the Duncan's multiple range test at the 5% confidence level.

Figure 3-3 Root Se content in *C. betacea* seedlings

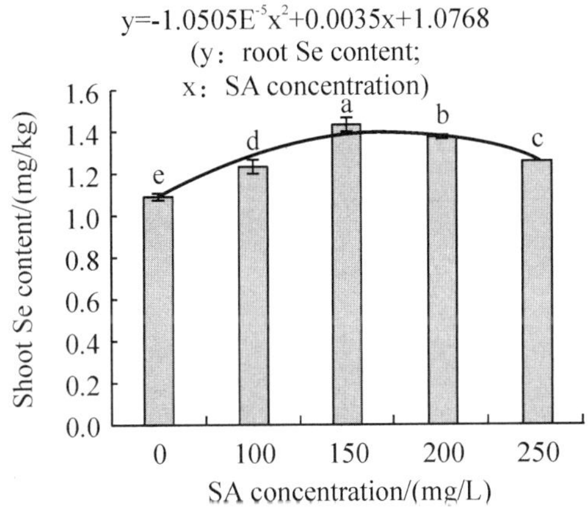

Note: Different lowercase letters within a column indicate significant differences based on one-way analysis of variance in SPSS 20. 0 followed by the Duncan's multiple range test at the 5% confidence level.

Figure 3-4 Shoot Se content in *C. betacea* seedlings

3. 6 Correlation and grey relational analyses

Correlation analysis was utilized to examine the relationships among the various parameters (Table 3-4). There were positive correlations observed among the different parameters. Additionally, grey relational analysis was used to assess the grey correlations of the different parameters with the shoot Se content (Figure 3-5). Grey relational analysis revealed that the grey correlation coefficients of all parameters with the shoot Se content ranging from 0. 21 to 0. 46. The top three grey correlation coefficients were the chlorophyll b content, net photosynthetic rate, and shoot biomass. These three parameters were found to be closely associated with the shoot Se content.

Table 3-4 Correlations among the different indicators

Indicator	RB	SB	Cha	Chb	Car	Tch	Pn	Gs	Ci	Tr	SOD	POD	CAT	SP	RSe	SSe
RB	1.000															
SB	0.853 * *	1.000														
Cha	0.858 * *	0.913 * *	1.000													
Chb	0.917 * *	0.908 * *	0.967 * *	1.000												
Car	0.880 * *	0.974 * *	0.913 * *	0.930 * *	1.000											
Tch	0.877 * *	0.917 * *	0.998 * *	0.981 * *	0.922 * *	1.000										
Pn	0.872 * *	0.947 * *	0.982 * *	0.959 * *	0.940 * *	0.982 * *	1.000									
Gs	0.856 * *	0.928 * *	0.973 * *	0.928 * *	0.911 * *	0.968 * *	0.991 * *	1.000								
Ci	0.717 * *	0.819 * *	0.767 * *	0.714 * *	0.833 * *	0.759 * *	0.782 * *	0.785 * *	1.000							
Tr	0.837 * *	0.944 * *	0.981 * *	0.939 * *	0.923 * *	0.977 * *	0.996 * *	0.992 * *	0.776 * *	1.000						
SOD	0.882 * *	0.917 * *	0.904 * *	0.900 * *	0.899 * *	0.909 * *	0.938 * *	0.923 * *	0.801 * *	0.926 * *	1.000					
POD	0.845 * *	0.981 * *	0.926 * *	0.900 * *	0.964 * *	0.925 * *	0.968 * *	0.958 * *	0.838 * *	0.967 * *	0.928 * *	1.000				
CAT	0.734 * *	0.927 * *	0.747 * *	0.772 * *	0.935 * *	0.757 * *	0.814 * *	0.775 * *	0.766 * *	0.802 * *	0.792 * *	0.907 * *	1.000			
SP	0.864 * *	0.858 * *	0.796 * *	0.851 * *	0.921 * *	0.814 * *	0.844 * *	0.800 * *	0.755 * *	0.808 * *	0.839 * *	0.874 * *	0.900 * *	1.000		
RSe	0.943 * *	0.805 * *	0.905 * *	0.948 * *	0.844 * *	0.921 * *	0.890 * *	0.873 * *	0.647 * *	0.860 * *	0.828 * *	0.801 * *	0.636 *	0.779 * *	1.000	
SSe	0.938 * *	0.963 * *	0.934 * *	0.955 * *	0.968 * *	0.944 * *	0.953 * *	0.937 * *	0.798 * *	0.935 * *	0.900 * *	0.950 * *	0.868 * *	0.890 * *	0.911 * *	1.000

Note: N = 15. * * =Correlation is significant at the 0.01 level (2-tailed test). * =Correlation is significant at the 0.05 level (2-tailed test). RB = root biomass, SB = shoot biomass Cha = chlorophyll a content, Chb = chlorophyll b content, Car = carotenoid content, Tch = total chlorophyll content, Pn = net photosynthetic rate, Gs = stomatal conductance, Ci = intercellular CO_2 concentration, Tr = transpiration rate, SOD = SOD activity, POD = POD activity, CAT = CAT activity, SP = soluble protein content, RSe = root Se content, SSe = shoot Se content.

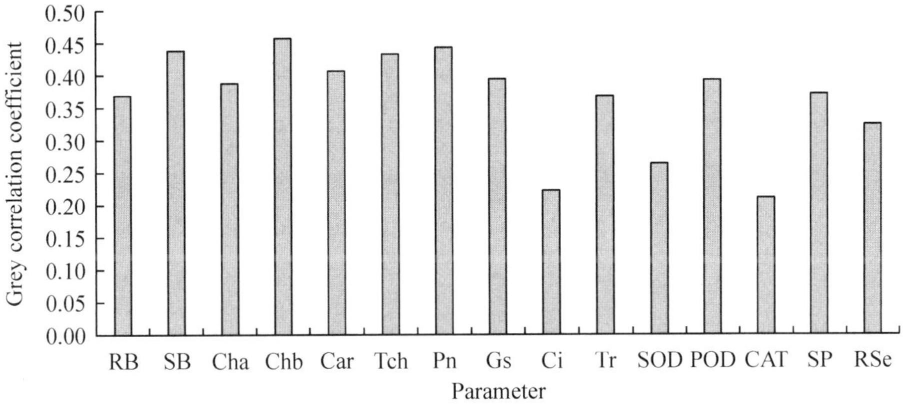

Note: RB = root biomass, SB = shoot biomass Cha = chlorophyll a content, Chb = chlorophyll b content, Car = carotenoid content, Tch = total chlorophyll content, Pn = net photosynthetic rate, Gs = stomatal conductance, Ci = intercellular CO_2 concentration, Tr = transpiration rate, SOD = SOD activity, POD = POD activity, CAT = CAT activity, SP = soluble protein content, RSe = root Se content.

Figure 3-5 Grey correlation coefficients of different parameters with the shoot Se content

4 Discussion

SA can enhance plant growth, as well as physiological and biochemical processes, by increasing soluble sugar, protein, antioxidant enzymes, leaf water content, membrane stability index, chlorophyll content, and chlorophyll stability index (Farhangi et al., 2018). With an increase in SA concentration, the primary agronomic traits of *Stevia rebaudiana* initially increase and then decrease, and applying an appropriate concentration of SA to *S. rebaudiana* can promote its growth and yield (Chen et al., 2019). Additionally, SA has been shown to promote tomato growth at a concentration of 100 μmol/L; however, at a concentration of 200 μmol/L, it inhibits tomato growth (Wang et al., 2019). Furthermore, under Cd or Se stress, SA has been shown to promote crop growth by increasing biomass (Tang et al., 2023; Liu et al., 2021; Li et al., 2022). In this experiment, SA increased the biomass of *C. betacea* seedlings under Se stress, with a concentration of 150 mg/L resulting in the highest biomass. These findings align with previous studies (Tang et al., 2023; Liu et al., 2021; Li et al., 2022), indicating that SA can promote the growth of *C. betacea* under Se stress. This may be attributed to SA enhancing photosynthesis and improving stomatal characteristics, ultimately increasing plant biomass by converting water and CO_2 into essential food raw materials (Fayez et al., 2014; Yusuf et al., 2008).

SA can induce the expression of photosynthetic-related genes and the activity of photosynthetic-related enzymes by triggering the reactive oxygen species (ROS), other signals, and photosynthetic factors such as photosynthetic pigment content, photosynthetic organ performance, and carbon assimilation efficiency (Tang et al., 2015). Photosynthesis is a vital plant physiological response, and the level of photosynthetic pigment content reflects the intensity of photosynthesis, playing a crucial role in plant biomass accumulation and productivity (Huang et al., 2018; Arfan et al., 2007). SA can preserve the stability of the photosynthetic system, leading to better leaf photosynthetic performance (Wang et al., 2020). When SA was applied to tobacco under triclosan stress, the changes in photosynthetic pigment content remained consistent with changes in biomass, with the content increasing along with plant growth (Guan et al., 2020). In this study, SA was found to increase the levels of chlorophyll a, chlorophyll b, total chlorophyll, and carotenoid in *C. betacea* seedlings under Se stress, with the optimal concentration being 150 mg/L. This outcome aligns with the biomass performance of *C. betacea* seedlings and previous research findings (Farhangi-Abriz et al., 2018; Guan et al., 2020), suggesting that SA can enhance dry matter production by improving the photosynthetic capacity of *C. betacea* seedlings. This is likely due to SA acting as an antioxidant, reducing H_2O_2 in the chloroplast, and increasing total chlorophyll content, thereby stimulating plant growth (Dawood et al., 2012).

The increase in the net photosynthetic rate is reflected in the increase in dry matter (Fariduddin et al., 2003). A study has found that spraying SA increased the net photosynthetic rate of corn (Gramineae), soybean (Leguminosae), and rapeseed (Cruciferae) (Khan et al., 2003; Dawood et al., 2012). These three plants belong to different families, so it is inferred that SA might promote the net photosynthetic rate of most plants. The results of this experiment are consistent with previous studies (Khan et al., 2003; Dawood et al., 2012). After spraying different concentrations of SA, the net photosynthetic rate and intercellular carbon dioxide concentration of *C. betacea* seedlings were higher than the control. Transpiration is an essential mechanism for nutrient absorption through a water potential gradient, which can drive nutrients to the roots and upwards to the shoots (Guo et al., 2009). Accumulating evidence suggests that SA treated with high transpiration (Shi et al., 2009). SA treatment can increase the transpiration and photosynthesis rate of perennial ryegrass, thereby reducing the adverse effects of Cd toxicity stress (Wang et al., 2013). In the present study, transpiration first increased and then decreased with the increased SA concentration under Se stress. This might be attributed to the fact that proper SA concentration can promote plant photosynthesis, while high SA concentration causes stomata to close and limits the Calvin cycle (Janda et al., 2012). SA can enhance ABA (growth hormone), which biosynthesizes various anti-stress proteins, enhances the antioxidant enzyme system, reduces ROS production, maintains membrane integrity, and regulates stomatal conductance. These processes ensure that mesophyll cells can use more carbon dioxide, thereby increasing the photosynthesis and transpiration rate (Faried et al., 2017).

Plants possess highly efficient, complex enzyme and non-enzymatic antioxidant defense systems that enable them to combat the harmful effects of free radicals(Dave, 2009). SA has been found to alleviate stress in tomatoes by enhancing antioxidant enzyme activity and gene expression (Chanda et al., 2009). In environments with low potassium stress, SA plays a crucial role in eliminating ROS through the antioxidant enzyme system and intricate post-transcriptional level regulation pathway, while an optimal concentration of SA can boost antioxidant enzyme activities and soluble protein content in plants, but excessive amounts of SA can prove toxic to different plant species due to varying levels of sensitivity (Xu et al., 2016). In this study, it is observed that application of SA increased the activities of SOD, CAT, and POD of *C. betacea* seedlings under Se stress, as well as the soluble protein content, aligning with previous research findings (Chanda et al., 2009; Xu et al., 2016). The previous study shows that the appropriate concentration of Se also influences the activities of CAT, glutathione peroxidase (GPX), and SOD, with Se triggering post-translational oxidative modifications of proteins (Talbi et al., 2018). It can be conjectured that there may be a synergistic effect between SA and Se in enhancing antioxidant enzyme activity and soluble protein content. This possible synergy could be explained by SA's role in facilitating Se absorption, leading to enzyme oxidative modifications and regulation of SA levels.

Although Se is widely distributed in the soil, its content is very low, and the active Se (water-soluble Se and exchangeable Se) content in plants is comparatively lower. The Se content in crops is closely related to the Se content of the soil, and the soil with high effective state content also has high Se content in plants (An, 2017). When the soil Se concentration is lower than 25 mg/kg, the amount of Se extraction by the shoots of *C. betacea* seedlings tends to increase. However, when the concentration is higher than 25 mg/kg, the amount of Se extraction by shoots decreases (Lin et al., 2020). Additionally, SA added to the nutrient solution can be considered as a bio-stimulating compound that can increase the absorption efficiency of Se by lettuce (roots and leaves)and lead to increased Se accumulation in leaves

(Smoleń et al., 2019). Under Se stress, SA has various effects on Se absorption in different crops (Li et al., 2022; Mostofa et al., 2020; Smoleń et al., 2018). In this study, SA increased the Se content in different organs of *C. betacea* seedlings, with the highest Se content found at 150 mg/L SA. The root Se content showed a linear relationship with the SA concentration, while the shoot Se content exhibited a quadratic polynomial relationship with the SA concentration. These results are consistent with previous studies (Li et al., 2022; Smoleń et al., 2019), indicating that SA could promote Se absorption in *C. betacea* under Se stress. Additionally, the content of root Se and shoot Se was positively correlated with other parameters in this study, and grey relational analysis showed that the chlorophyll b content, net photosynthetic rate, and shoot biomass were most closely related to the shoot Se content. These results suggest that growth and Se absorption promote each other in *C. betacea* under Se stress. Further study is needed to determine the mechanisms behind these results.

5 Conclusion

Under Se stress, the different concentrations of SA promoted the growth of *C. betacea* seedlings by increasing their biomass, photosynthetic pigment content, and antioxidant enzyme activity. Additionally, SA enhanced the Se absorption in various organs of *C. betacea* seedlings. The optimal concentration of SA was found to be 150 mg/L for growth and Se absorption of *C. betacea* seedlings. Future research should delve into the mechanism behind *C. betacea*'s Se absorption facilitated by SA.

References

[1] AN M Y. The main factors affecting the transformation of selenium forms in selenium-rich soils and their effects on crop absorption[D]. Fujian Agriculture and Forestry University, Fuzhou, China, 2017.

[2] ARFAN M, ATHAR H R, ASHRAF M. Does exogenous application of salicylic acid through the rooting medium modulate growth and photosynthetic capacity in two differently adapted spring wheat cultivars under salt stress[J]? Journal of Plant Physiology, 2007, 164 (6):685-694.

[3] BAO S D. Soil agro-chemistrical analysis[M]. China Agriculture Press, Beijing,

China, 2000.

[4] CHANDA S, DAVE R. In vitro models for antioxidant activity evaluation and some medicinal plants possessing antioxidant properties: An overview [J]. African Journal of Microbiology Research, 2009, 3(13): 981-996.

[5] CHEN J T, YI B, CHEN A M, et al. Effects of spraying salicylic acid at seedling stage on main agronomic characteristics and glycoside content of stevia [J]. Acta Botanica Boreali-Occidentalia Sinica, 2019, 39(1): 149-155.

[6] CUTT J R, KLESSIG D F. Pathogenesis-related proteins/genes involved in plant defense. Springer, Vienna, 1992.

[7] DAVE R. In vitro models for antioxidant activity evaluation and some medicinal plants possessing antioxidant properties: An overview [J]. African Journal of Microbiology Research, 2009, 3(13): 981-996.

[8] DAWOOD M, SADAK M S, HOZAYEN M. Physiological role of salicylic acid in improving performance, yield and some biochemical aspects of sunflower plant grown under newly reclaimed sandy soil [J]. Australian Journal of Basic and Applied Sciences, 2012, 6(4): 82-89.

[9] EL-RAMADY H, ABDALLA N, TAHA H S, et al. Selenium and nano-selenium in plant nutrition [J]. Environmental Chemistry Letters, 2016, 14(1): 123-147.

[10] FARHANGI-ABRIZ S, GHASSEMI-GOLEZANI K. How can salicylic acid and jasmonic acid mitigate salt toxicity in soybean plants? [J]. Ecotoxicology and Environmental Safety, 2018, 147: 1010-1016.

[11] FARIDUDDIN Q, HAYAT S, AHMAD A. Salicylic acid influences net photosynthetic rate, carboxylation efficiency, nitrate reductase activity, and seed yield in *Brassica juncea* [J]. Photosynthetica, 2003, 41(2): 281-284.

[12] FARIED H N, AYYUB C M, AMJAD M, et al. Salicylic acid confers salt tolerance in potato plants by improving water relations, gaseous exchange, antioxidant activities and osmoregulation [J]. Journal of the Science of Food and Agriculture, 2017, 97(6): 1868-1875.

[13] FAYEZ K A, BAZAID S A. Improving drought and salinity tolerance in barley by application of salicylic acid and potassium nitrate [J]. Journal of the Saudi Society of Agricultural Sciences, 2014, 13(1): 45-55.

[14] FENG J, XING Y, ZHANG X, et al. Effects of salicylic acid, abscisic acid and seaweed extract on drought resistance and nutrient uptake of apple seedlings under drought

stress[J]. China Fruits, 2019, (6):42-48.

[15] GUAN C F, WANG C, WU H, et al. Salicylic acid application alleviates the adverse effects of triclosan stress in tobacco plants through the improvement of plant photosynthesis and enhancing antioxidant system[J]. Environmental Science and Pollution Research, 2020, 27(2):1359-1372.

[16] GUO B, LIANG Y, ZHU Y. Does salicylic acid regulate antioxidant defense system, cell death, cadmium uptake and partitioning to acquire cadmium tolerance in rice[J]? Journal of Plant Physiology, 2009, 166(1):20-31.

[17] HAO Z B, CANG J, XU Z. Plant physiology experiment[M]. Harbin Institute of Technology Press, Harbin, China, 2024.

[18] HUAN Y, YANG L, LIU Q, et al. Effects of indole acetic acid on the growth and selenium absorption characteristics of *Cyphomandra betacea* seedlings[J]. Acta Physiologiae Plantarum, 2021, 43(5):1-8.

[19] HUANG J J, LIN L J, CHEN F B, et al. Effects of intercropping *Bidens* L. species on growth and cadmium accumulation of grape seedlings[J]. Journal of Sichuan Agricultural University, 2018, 36(4):481-487.

[20] JANDA K, HIDEG É, SZALAI G, et al. Salicylic acid may indirectly influence the photosynthetic electron transport[J]. Journal of Plant Physiology, 2012, 169(10):971-978.

[21] KHAN W, PRITHIVIRAJ B, SMITH D L. Photosynthetic responses of corn and soybean to foliar application of salicylates[J]. Journal of Plant Physiology, 2003, 160(5):485-492.

[22] LI Z, FAN R, PENG X, et al. Salicylic acid alleviates selenium stress and promotes selenium uptake of grapevine[J]. Physiology and Molecular Biology of Plants, 2022, 28(3):625-635.

[23] LIN L, LI Z, WANG J, et al. 24-epibrassinolide promotes selenium uptake in grapevine under selenium stress[J]. Scientia Horticulturae, 2023, 308:111564.

[24] LIN L, SUN J, CUI T, et al. Selenium accumulation characteristics of *Cyphomandra betacea* (*Solanum betaceum*) seedlings[J]. Physiology and Molecular Biology of Plants, 2020, 26(7):1375-1383.

[25] LIU L, LI X, HUANG K, et al. Salicylic acid promotes growth and affects cadmium accumulation of *Cyphomandra betacea* seedlings[J]. Environ Prog Sustainable Energy, 2021, 41(3):e13787.

[26] MORENO-ESCAMILLA J O, ALVAREZ-PARRILLA E, DE LA ROSA L A, et al. Effect of different elicitors and preharvest day application on the content of phytochemicals and antioxidant activity of butterhead lettuce (*Lactuca sativa* var. *capitata*) produced under hydroponic conditions [J]. Journal of Agricultural and Food Chemistry, 2017, 65 (26): 5244-5254.

[27] MOSTOFA M G, RAHMAN M M, SIDDIQUI M N, et al. Salicylic acid antagonizes selenium phytotoxicity in rice: selenium homeostasis, oxidative stress metabolism and methylglyoxal detoxification[J]. Journal of Hazardous Materials, 2020, 394: 122572.

[28] NAPOLEÃO T A, SOARES G, VITAL C E, et al. Methyl jasmonate and salicylic acid are able to modify cell wall but only salicylic acid alters biomass digestibility in the model grass *Brachypodium distachyon*[J]. Plant Science, 2017, 263: 46-54.

[29] PAN S, LU R, LI H, et al. Mutual intercropping affects selenium uptake of eggplant seedlings[J]. International Journal of Environmental and Analytical Chemistry, 2021, 101 (15): 2866-2875.

[30] RASKIN I. Role of salicylic acid in plants[J]. Annual Review of Plant Physiology and Plant Molecular Biology, 1992, 43(1): 439-463.

[31] REIS A R D, EI-RAMADY H, SANTOS E F, et al. Overview of selenium deficiency and toxicity worldwide: affected areas, selenium-related health issues, and case studies[J]. Selenium in Plants, 2017, 11: 209-230.

[32] SAINI R K, PRASHANTH K V H, SHETTY N P, et al. Elicitors, SA and MJ enhance carotenoids and tocopherol biosynthesis and expression of antioxidant related genes in *Moringa oleifera* Lam. leaves[J]. Acta Physiologiae Plantarum, 2014, 36(10): 2695-2704.

[33] SHI G R, CAI Q S, LIU Q Q, et al. Salicylic acid-mediated alleviation of cadmium toxicity in hemp plants in relation to cadmium uptake, photosynthesis, and antioxidant enzymes[J]. Acta Physiologiae Plantarum, 2009, 31(5): 969-977.

[34] SMOLEń S, KOWALSKA I, KOVáCIK P, et al. Biofortification of six varieties of lettuce (*Lactuca sativa* L.) with iodine and selenium in combination with the application of salicylic acid[J]. Frontiers in Plant Science, 2019, 10: 143.

[35] SMOLEń S, KOWALSKA I, SKOCZYLAS Ł, et al. The effect of salicylic acid on biofortification with iodine and selenium and the quality of potato cultivated in the NFT system[J]. Scientia Horticulturae, 2018, 240: 530-543.

[36] SUN X, HUANG W, LI B. Effects of exogenous salicylic acid on physiological and biochemical indexes and related gene expression in *Platycodon grandiflorus* under drought

stress[J]. Journal of Agricultural Science and Technology, 2022, 24(1):63-70.

[37] TALBI W, GHAZOUANI T, BRACONI D, et al. Effects of selenium on oxidative damage and antioxidant enzymes of eukaryotic cells: wine *Saccharomyces cerevisiae*[J]. Journal of Applied Microbiology, 2018, 126(2):555-566.

[38] TANG W, LIANG L, XIE Y, et al. Foliar application of salicylic acid inhibits the cadmium uptake and accumulation in lettuce (*Lactuca sativa* L.)[J]. Frontiers in Plant Science, 2023, 14:1200106.

[39] TANG Y P, WEN T, SUN X, et al. Research progress on the effects of salicylic acid on plant photosynthesis[J]. Acta Botanica Boreali-Occidentalia Sinica, 2015, 35(8):1701-1708.

[40] WANG Q H, LIANG X, DONG Y J, et al. Effects of exogenous salicylic acid and nitric oxide on physiological characteristics of perennial ryegrass under cadmium stress[J]. Journal of Plant Growth Regulation, 2013, 32(4):721-731.

[41] WANG W L, WANG X, ZHANG J, et al. Salicylic acid and cold priming induce late-spring freezing tolerance by maintaining cellular redox homeostasis and protecting photosynthetic apparatus in wheat[J]. Plant Growth Regulation, 2020, 90(1):109-121.

[42] WANG X H, GUO J K, JIA H L, et al. Exogenous salicylic acid alleviates the toxicity of cadmium to tomato[J]. Journal of Agro-Environment Science, 2019, 38(12):2705-2714.

[43] WINKEL L H E, VRIENS B, JONES G D, et al. Selenium cycling across soil-plant-atmosphere interfaces: a critical review[J]. Nutrients, 2015, 7(6):4199-4239.

[44] XU L, LI J, SONG W Q, et al. Salicylic acid alleviates tomato low potassium stress by increasing antioxidant enzyme activity and gene expression[J]. Acta Agriculturae Boreali-Sinica, 2016, 31(1):96-101.

[45] YANG L. Thinking on the present situation and development countermeasures of *Cyphomandra betacea* production in Tengchong City, Yunnan, China[J]. South China Agriculture. 2019, 13(5):113-114.

[46] YUSUF M, HASAN S A, ALI B, et al. Effect of salicylic acid on salinity-induced changes in *Brassica juncea*[J]. Journal of Integrative Plant Biology, 2008, 50:1096-1102.

[47] ZHANG D, HE N, YANG X, et al. Research advance on the important role of selenoprotein in human health[J]. Chinese Science Bulletin, 2022, 67(6):473-480.

[48] ZHENG Y, ZHANG R, ZHU Y, et al. Salicylic acid improves *Nasturtium officinale* phytoremediation capability for cadmium-contaminated paddy soils[J]. Frontiers in Plant Science, 2022, 13:1059175.

Section Ⅱ Uniconazole

1 Introduction

Selenium (Se) is a nutrient that is essential to maintain human body functions (Yin et al., 2019). However, owing to the excessive application of exogenous Se fertilizer or the improper application of Se, a large amount of Se is released into the environment (Zhu et al., 2021). Se has gradually become a new emerging potential pollutant around the world (Winkel et al., 2012). Enshi of Hubei Province, Kaiyang of Guizhou Province and Ziyang of Shaanxi Province are all typical areas in China that are rich in Se (Jiang et al., 2007), and poisoning incidents caused by Se pollution occur on occasion (Fang et al., 2004). The excessive Se in soil also inhibits the growth of plants (Van Hoewyk, 2013; Sors et al., 2005). Thus, finding appropriate methods to alleviate the toxic effects of Se pollution (Se stress) on plant growth is necessary.

Studies have shown that plant growth regulators enhance the nutrient absorption capacity of roots, while promote crop growth, and increase production (Gong et al., 2019; Yuan, 2013). Uniconazole (S3307) is a triazole plant growth regulator with multiple effects. It regulates plant shape, controls upper and lower levels of height, enhances stress (drought, high temperature, salt damage) resistance, and increases yield (Lu et al., 2017). In plants, S3307 affects the absorption, transportation, and utilization of soil nutrients; however, this effect is related to the plant species (Li et al., 2012; David et al., 2000; Wu et al., 2009). Under different nitrogen fertilizer levels, S3307 promotes nitrogen absorption, assimilation, and metabolism in soybean roots (Wan et al., 2012). S3307 also regulates the distribution of nutrient elements, increases the nitrogen, phosphorus, and potassium content in the root tuber, and increases the harvest index of nitrogen, phosphorus, and potassium fertilizers in sweet potato plants (Xie et al., 2020), and the same regulation occurres in potatoes (Li et al., 2012). S3307 also promotes the absorption of medium and trace elements such as calcium, magnesium, iron, copper, zinc, and manganese by plants (Li et al., 2012; El-Bahay et al., 2003; Kandil et al., 2008; Zhang et al., 2020). Furthermore, studies have proven that S3307 promotes the absorption of nutrients by changing the internal structure of plant roots (Meng et al., 2018), increasing the root vitality (Wan et al., 2012; Meng et al., 2018), and comprehensively changing various physiological and biochemical processes (Liu et al., 2015; Huang et al., 2015; Wei et al., 2017). However, the effects of S3307 on the Se uptake of plants are unclear.

Cyphomandra betacea is a fruit tree rich in proteins and vitamins with high nutritional value (Hassan et al., 2013; Kahia et al., 2015). A previous study reported that the high Se concentration in soil increased the Se content in *C. betacea* (Lin et al., 2020a). Therefore, further improving the Se uptake in *C. betacea* is beneficial to Se contaminated areas. In this study, the effects of S3307 on the growth and Se accumulation characteristics of *C. betacea* seedlings were studied to improve the Se accumulation ability of *C. betacea*. This study aimed to determine the optimal concentration of S3307 that promotes the growth and Se accumulation of *C. betacea* seedlings.

2 Materials and methods

2.1 Materials

The mature seeds of *C. betacea* were collected from Chengdu Campus of Sichuan Agricultural University (30°71 'N, 103°87 'E) in September 2020, air-dried, sown in pots filled with moist perlite, and incubated in a climate chamber under the conditions: 14-h days (light intensity of 4,000 Lux) at 25 ℃ with 70% relative humidity and 10-h night (0 Lux) at 20 ℃ with 90% relative humidity. (Li et al., 2022). The pots were watered with Hoagland's nutrient solution every 3 days.

2.2 Experimental design

In October 2020, four uniform *C. betacea* seedlings with 15 cm height were transplanted into each pot (15 cm height and 15 cm diameter) filled with sterile perlite. The seedlings were maintained under the aforementioned conditions and watered with the Hoagland nutrient solution (100 mL) every 3 days. One week later, different concentrations (0, 5, 10, 20, and 50 mg/L) of S3307 (Li et al., 2012; Kandil et al., 2008; Zhang et al., 2020) were sprayed onto the leaves of *C. betacea* seedlings (25 mL per treatment; until the solution has been dripping from the leaf edge); triplicates were maintained per treatment. After that, the plants were watered once every 3 days with 100 ml of Hoagland nutrient solution containing 0. 1 mg/L

Se (analytically pure Na_2SeO_3)(Li et al., 2022).

2.3 Sample analysis

One month after S3307 treatment (November 2020), the net photosynthetic rate (Pn), transpiration rate (Tr), stomatal conductance (Gs), and intercellular CO_2 concentration (Ci) of the functional leaves (sixth leaf from the top) were determined using an LI-6400 portable photosynthesis meter (LI-COR, USA) with a light intensity of 1, 200 μmol/m²/s, a temperature of 25 ℃, and a CO_2 concentration of 400 μmol/mol. The leaves from the same part of the plants were collected to determine the photosynthetic pigment (chlorophyll a, chlorophyll b, total chlorophyll, and carotenoid) content by ethanol acetone extraction method (Hao et al., 2004). The fifth leaf from the top of the plant was collected to determine the antioxidant enzyme [superoxide dismutase (SOD), peroxidase (POD), and catalase (CAT)] activities and soluble protein content. The SOD, POD, and CAT activities were determined by the nitroblue tetrazolium photoreduction method, guaiacol colorimetric method, and potassium permanganate titration method, respectively; the soluble protein content was determined by Coomassie brilliant blue staining. The details of determining the antioxidant enzyme activities and soluble protein content were described in previous studies (Hao et al., 2004; Lin et al., 2020b). After that, all plants was harvested and treated as reported by Lin et al. (2020b), and the root, stem, and leaf biomass (dry weight) was determined using an electronic balance. The plant was digested in $HNO_3/HClO_4$ and the Se content was determined using an iCAP 6300 ICP-MS spectrometer (Thermo Scientific, Waltham, MA, USA), with HCl as the reducing agent (Bao, 2000).

2.4 Statistical analyses

Data was subjected to a one-way analysis of variance (ANOVA) using the least significant difference test ($P < 0.05$). Polynomial regression analyses were also performed. The analyses were carried out in SPSS 20.0 (IBM, Chicago, IL, USA). The shoot biomass was determined as the sum of stem biomass and leaf biomass. The Se content in shoots was determined as follows: Se content in shoots = (Se content in stems × stem biomass + Se content in leaves × leaf biomass)/ shoot biomass (Rastmanesh et al., 2010).

3 Results

3.1 Biomass

At 5, 10, and 20 mg/L concentrations of S3307, the biomass of the different organs of *C. betacea* seedlings was higher than the control; 20 mg/L S3307 resulted in the maximum biomass (Table 3-5). Meanwhile, the stem, leaf, and shoot biomass was higher when treated with 50 mg/L S3307 than the control, while the root biomass did not change significantly. S3307 at 5, 10, and 20 mg/L increased the root biomass by 6.27%, 13.47%, and 19.47%, respectively, compared with the control and the shoot biomass by 12.57%, 25.14%, and 36.29%, respectively. Besides, the root biomass and shoot biomass showed a polynomial regression relationship with S3307 concentration as follows: For root biomass, $y = -2E^{-05}x^2 + 0.0012x + 0.0748$, $R^2 = 0.9959$ ($P < 0.01$); and for shoot biomass, $y = -0.0002x^2 + 0.0103x + 0.3498$, $R^2 = 0.9965$ ($P < 0.01$).

Table 3-5 Biomass of *C. betacea* seedlings

S3307 concentration/ (mg/L)	Root/ (g/plant)	Stem/ (g/plant)	Leaf/ (g/plant)	Shoot/ (g/plant)
0	0.0750±0.0013d	0.088±0.003d	0.262±0.004d	0.350±0.007d
5	0.0797±0.0001c	0.096±0.001c	0.298±0.008c	0.394±0.009c
10	0.0851±0.0022b	0.110±0.005b	0.328±0.010b	0.438±0.014b
20	0.0896±0.0024a	0.122±0.006a	0.354±0.009a	0.477±0.007a
50	0.0747±0.0020d	0.096±0.003c	0.284±0.015c	0.379±0.015c

Note: Different lowercase letters indicate significant differences among the treatments (least significant difference test, $P < 0.05$). Shoot biomass = stem biomass + leaf biomass.

3.2 Photosynthetic pigment content

S3307 at different concentrations increased the photosynthetic pigment content in *C. betacea* seedlings compared with the control (Table 3-6). Moreover, S3307 at 5, 10, 20, and 50 mg/L increased the total chlorophyll content by 30.41%, 53.30%, 61.36%, and 24.25%, respectively, and the carotenoid content by 31.33%, 119.28%, 78.92%, and 68.67%, respectively.

Table 3-6 Photosynthetic pigment content in *C. betacea* seedlings

S3307 concentration/ (mg/L)	Chlorophyll a/ (mg/g)	Chlorophyll b/ (mg/g)	Carotenoid/ (mg/g)	Total chlorophyll/ (mg/g)
0	0. 889±0. 023e	0. 216±0. 013d	0. 166±0. 006d	1. 105±0. 023e
5	1. 131±0. 024c	0. 310±0. 009c	0. 218±0. 007c	1. 441±0. 030c
10	1. 310±0. 028b	0. 384±0. 012b	0. 364±0. 012a	1. 694±0. 018b
20	1. 371±0. 016a	0. 412±0. 017a	0. 297±0. 017b	1. 783±0. 009a
50	1. 060±0. 021d	0. 313±0. 015c	0. 280±0. 014b	1. 373±0. 034d

Note: Different lowercase letters indicate significant differences among the treatments (least significant difference test, $P < 0.05$).

3.3 Photosynthetic characteristics

Furthermore, S3307 at different concentrations increased the Pn, Gs, Ci, and Tr of *C. betacea* seedlings (Table 3-7). S3307 at 5, 10, 20, and 50 mg/L increased the Pn by 46. 89%, 48. 24%, 50. 50%, and 5. 07%, respectively; the Gs by 125. 93%, 185. 75%, 201. 99%, and 23. 65%, respectively, the Ci by 47. 84%, 63. 99%, 66. 16% and 22. 58%, respectively; and the Tr by 105. 57%, 152. 86%, 172. 29%, and 22. 29%, respectively, compared with the control.

Table 3-7 Photosynthetic characteristics of *C. betacea* seedlings

S3307 concentration/ (mg/L)	Pn/ [μmol CO_2/(m^2/s)]	Gs/ [mol H_2O/(m^2/s)]	Ci/ (μmol CO_2/mol)	Tr/ [mmol H_2O/(m^2/s)]
0	4. 988±0. 105c	0. 0351±0. 0007e	148. 871±2. 035d	0. 664±0. 006e
5	7. 327±0. 119a	0. 0793±0. 0014c	220. 086±3. 982b	1. 365±0. 015c
10	7. 394±0. 134a	0. 1003±0. 0018b	244. 127±3. 695a	1. 679±0. 011b
20	7. 507±0. 090a	0. 1060±0. 0019a	247. 368±8. 014a	1. 808±0. 024a
50	5. 241±0. 068b	0. 0434±0. 0005d	182. 485±3. 402c	0. 812±0. 011d

Note: Different lowercase letters indicate significant differences among the treatments (least significant difference test, $P < 0.05$). Pn = net photosynthetic rate; Tr = transpiration rate; Gs = stomatal conductance; Ci = intercellular CO_2 concentration.

3.4 Antioxidant enzyme activity and soluble protein content

S3307 at 5 mg/L had no significant effect on the SOD activity of *C. betacea* seedlings compared with the control, while that at 10, 20, and 50 mg/L enhanced the SOD activity by

18.81%, 25.27%, and 8.29%, respectively (Table 3-8). Meanwhile, S3307 at different concentrations improved the CAT and POD activities of *C. betacea* seedlings compared with the control, with the maximum effect at 20 mg/L. S3307 at 5, 10, and 50 mg/L resulted in higher soluble protein content than that of the control, while at 20 mg/L lowered the soluble protein content.

Table 3-8 Antioxidant enzyme activity and soluble protein content in *C. betacea* seedlings

S3307 concentration/ (mg/L)	SOD activity/ (U/g)	CAT activity/ [mg/(g/min)]	POD activity/ [U/(g/min)]	Soluble protein content/ (mg/g)
0	219.6±3.42c	5.496±0.024e	1,201±17.56c	152.7±6.64c
5	230.1±8.71bc	6.467±0.094c	1,357±23.25b	243.4±7.41a
10	260.9±8.98a	6.696±0.063b	1,391±51.97b	232.4±9.58a
20	275.1±9.65a	6.976±0.105a	1,473±36.06a	116.6±5.87d
50	237.8±8.72b	5.796±0.089d	1,357±29.96b	207.2±8.80b

Note: Different lowercase letters indicate significant differences among the treatments (least significant difference test, $P < 0.05$).

3.5 Se content

S3307 at 5, 10, and 20 mg/L concentrations increased the Se content in the roots of *C. betacea* seedlings by 9.15%, 8.67%, and 12.58%, respectively, compared with the control, while 50 mg/L of S3307 had no significant effect (Table 3-9). S3307 increased the Se content of the stems, leaves, and shoots of *C. betacea* seedlings, compared with the control, with the maximum effect at 20 mg/L. S3307 at 5, 10, 20, and 50 mg/L increased the Se content of the shoots by 38.23%, 56.37%, 75.55%, and 26.75%, respectively, compared with the control. In addition, the Se content of the roots and shoots showed a polynomial regression relationship with the S3307 concentration as follows: For Se content in roots, $y = -0.0033x^2 + 0.162x + 16.897$, $R^2 = 0.8949$ ($0.01 \leqslant P < 0.05$) and for Se content in shoots, $y = -0.001x^2 + 0.0516x + 0.9166$, $R^2 = 0.9729$ ($P < 0.01$).

Table 3-9 Se content in *C. betacea* seedlings

S3307 concentration/ (mg/L)	Root/ (mg/kg)	Stem/ (mg/kg)	Leaf/ (mg/kg)	Shoot/ (mg/kg)
0	16.61±0.231c	0.946±0.018d	0.846±0.043e	0.871±0.036e
5	18.13±0.190b	1.399±0.022b	1.141±0.015c	1.204±0.008c
10	18.05±0.182b	1.424±0.034b	1.340±0.052b	1.362±0.037b

continued table

S3307 concentration/ (mg/L)	Root/ (mg/kg)	Stem/ (mg/kg)	Leaf/ (mg/kg)	Shoot/ (mg/kg)
20	18. 70±0. 219a	1. 682±0. 024a	1. 476±0. 044a	1. 529±0. 037a
50	16. 66±0. 240c	1. 211±0. 035c	1. 067±0. 035d	1. 104±0. 034d

Note: Different lowercase letters indicate significant differences among the treatments (least significant difference test, $P < 0.05$); Se content in shoot = (Se content in stem × stem biomass + Se content in leaf × leaf biomass)/ shoot biomass.

4 Discussion

Plant growth regulators have been widely used to regulate growth and increase crop yields (Pei et al., 2015; Jaleel et al., 2007). S3307, a member of the plant growth regulator family, is often used as a growth retardant due to its ability to regulate plant shape (Lu, et al. 2017). Studies have shown that S3307 inhibits plant growth; it significantly affects dwarfing, maintains crop growth, and improves crop lodging resistance (Gao et al., 2016; Ma et al., 2016; Jiang et al., 2010). However, not all S3307 has an inhibitory effect on plant growth. Within a certain concentration range, the higher the concentration of S3307, the greater the growth promotion effect. Beyond this concentration range, the promotion effect was weakened (He et al., 2017). These findings indicated that the mechanisms via which S3307 influences plant growth are complicated. Plants respond to S3307 treatment through various physiological and biochemical processes, such as the up-regulation expression of levels of endogenous hormones (cytokinin, abscisic acid, gibberellins, and indole acetic acid) (Liu et al., 2015; Huang et al., 2015; Wei et al., 2017). In this experiment, different concentrations of S3307 promoted the growth of *C. betacea* seedlings by increasing the biomass, and S3307 at 20 mg/L had the maximum biomass. Besides, the root and shoot biomass showed a polynomial regression relationship with S3307 concentration, consistent with the previous studies (He et al., 2017), indicating the effect of S3307 on plant growth has a dual nature. Therefore, S3307 at an optimal concentration could effectively promote the growth of *C. betacea* seedlings, and the present study proved 20 mg/L as the optimal concentrations.

In plants, S3307 promotes chlorophyll content; however, with the increase in S3307 concentration, the promotion effect gradually decreases and even resultes in an inhibitory effect (Gao et al., 2017; Liu et al., 2017; Zhong et al., 2015). In this experiment, the chlorophyll content in *C. betacea* seedlings increased first; it then decreased with the increase in S3307 concentration, consistent with previous studies (Gao et al., 2017; Liu et al., 2017; Zhong et al., 2015), indicating that the effect of S3307 on plant chloroplast content is concentration dependent. Several studies have shown that S3307 promotes cytokinin biosynthesis, which in turn promotes chloroplast differentiation and biosynthesis (Liu et al., 2015). In this experiment, S3307 increased the chlorophyll content in *C. betacea* seedlings, probably by promoting cytokinin biosynthesis (Liu et al., 2015). Application of S3307 (20 mg/L) significantly increases the Pn, Gs, Ci, and Tr of potatoes (Xu et al., 2009) and the Pn, Tr, and Gs of kidney beans (Wang et al., 2018). Similarly, in this experiment, S3307 at different concentrations increased the Pn, Gs, Ci, and Tr of *C. betacea* seedlings, indicating that S3307 promoted photosynthesis and transpiration, which is consistent with the previous studies (Xu et al., 2009; Wang et al., 2018). Therefore, applying S3307 at an optimal concentration could effectively promote photosynthesis and transpiration and improve the synthesis, absorption, and transportation of organic matter, thereby promoting the growth of *C. betacea* seedlings.

Furthermore, S3307 increases the activities of SOD, CAT, and POD in plants under a variety of adversities, thereby increasing stress resistance (El-Bahay et al., 2003; Kandil et al., 2008). In semi-arid areas, an appropriate concentration of S3307 significantly increases wheat's SOD, POD, and CAT activities (Ahmad et al., 2019). Under water-stressed conditions, S3307 increases the activities of SOD and POD in soybean leaves (Zhang et al., 2008). In this experiment, different concentrations of S3307 increased the SOD, CAT, and POD activities of *C. betacea* seedlings to some extent. Meanwhile, in this experiment, S3307 at 5, 10, and 50 mg/L increased the soluble protein content in *C. betacea* seedlings, while 20 mg/L of S3307 decreased, contradictory to the previous study (Ahmad et al., 2019; Zhang et al., 2008). The soluble protein performs an osmotic adjustment function in the plant body, significantly improving the water retention capacity and providing drought and other stress tolerance benefits (Geng, 2019; Li et al., 2008). Thus, the present observation indicates that S3307 is not the main factor affecting the soluble protein content.

Studies have reported that S3307 promotes the absorption of nutrients, such as nitrogen, phosphorus, potassium, calcium, magnesium, iron, copper, zinc, and manganese (Li et al., 2012; Wan et al., 2012; Xie et al., 2020; El-Bahay et al., 2003; Kandil et al., 2008; Zhang et al., 2020). The application of S3307 increases the trace element content of soybean and sunflower seedlings (iron and copper) (El-Bahay et al., 2003) and *Ammi visnaga* (iron, zinc, manganese, and copper)(Kandil et al., 2008). Meanwhile, different concentrations of S3307 (< 125 mg/L) increases the cadmium content of the roots and shoots of farmland and mine ecology type *Bidens pilosa*, a hyperaccumulator (Huang et al., 2019). In this study, different concentrations of S3307 increased the Se content in various organs of *C. betacea* seedlings to some extent, and the maximum effect was observed at 20 mg/L, consistent with the previous studies (Huang et al., 2019). In addition, the Se content of roots and shoots showed a polynomial regression relationship with the S3307 concentration. These results indicate that S3307 promotes the Se uptake in *C. betacea*. When the Se content in plants is in excess, it causes toxic effects and inhibits growth (Wang et al., 2021).

5 Conclusion

S3307 promotes the growth of *C. betacea* seedlings by increasing the biomass, photosynthesis, and antioxidant enzyme activity. S3307 also increases the Se uptake in *C. betacea* seedlings, and the Se content showed a significant positive correlation with the biomass. Therefore, S3307 could be used in Se-enriched *C. betacea* seedlings, and 20 mg/L is the optimal concentration.

References

[1] AHMAD I, KAMRAN M, YANG X N, ct al. Effccts of applying uniconazole alone or combined with manganese on the photosynthetic efficiency, antioxidant defense system, and yield in wheat in semiarid regions [J]. Agricultural Water Management, 2019, 216 (C): 400-414.

[2] BAO S D. Soil Chemical Analysis [M]. Agriculture Press, Beijing, China, 2000.

[3] DAVID A B, ZHANG G P, CHEN J X. Effect of timing of N fertilizer and PGR on N uptake and utilization in wheat [J]. Journal of Zhejiang University (Agriculture and Life Sciences), 2000, 26(2): 159-164.

[4] EL-BAHAY M M, DOWIDAR A E, ABBAS S M. Physiological responses of foliar applications of uniconazole on soybean and sunflower plants [J]. Egyptian Journal of Biotechnology, 2003(13): 229-253.

[5] FANG W, WU P. Elevated selenium and other mineral element concentrations in soil

and plant tissue in bone coal sites inHaoping area, Ziyang County, China[J]. Plant and Soil, 2004, 261(1-2):135-146.

[6]GAO J Q, CHEN S, PENG Q, et al. Effect of uniconazole application on seedling growth and seed yield and quality in *Brassica napus* L[J]. Jiangsu Journal of Agricultural Sciences, 2016, 32(2):305-312.

[7]GAO Y, WANG J, SHI L J, et al. Effect of uniconazole on the lodging-resistance character and photosynthetic pigment of foxtail millet[J]. Journal of Shanxi Agricultural Sciences, 2017, 45(8):1232-1236.

[8]GENG Y H. Study on resistance of woody plants under drought stress[J]. Journal of Shandong Agricultural University (Natural Science Edition), 2019, 49(1):12-18.

[9]GONG L, LI Z B, ZENG E, et al. Effects of liquid fertilizers with plant growth regulators on nutrient absorption and fertilizer using efficiency of sweet corn[J]. Guangdong Agricultural Sciences, 2019, 46(11):54-61.

[10]HAO Z B, CANG J, XU Z. Plant physiology experiment[M]. Harbin Institute of Technology Press, Harbin, China, 2004.

[11]HASSAN S H A, BAKAR M F A. Antioxidative and anticholinesterase activity of *Cyphomandra betacea* fruit[J]. The Scientific World Journal, 2013, 2013(8):278071.

[12]HE J, LIN L J, MA Q Q, et al. Uniconazole (S-3307) strengthens the growth and cadmium accumulation of accumulator plant *Malachium aquaticum*[J]. International Journal of Phytoremediation, 2017, 19(4):348-352.

[13]HUAN Y M, YANG L, LIU Q, et al. Effects of indole acetic acid on the growth and selenium absorption characteristics of *Cyphomandra betacea* seedlings[J]. Acta Physiologiae Plantarum, 2021, 43(5):1-8.

[14]HUANG K W, SHAN S Y, LU C Y, et al. Effects of spraying uniconazole (S330) on cadmium accumulation of two ecotypes of *Bidens pilosa*[J]. Journal of Ecology and Rural Environment, 2019, 35(9):1190-1196.

[15]HUANG M J, YANG F, LIU Y, et al. Using proteomic analysis to investigate uniconazole-induced phytohormone variation and starch accumulation in duckweed (*Landoltia punctata*)[J]. BMC Biotechnology, 2015, 15(1):1-13.

[16]JALEEL C A, KISHOREKUMAR A, MANIVANNAN P, et al. Alterations in carbohydrate metabolism and enhancement in tuber production in white yam (*Dioscorea rotundata* Poir.) under triadimefon and hexaconazole applications[J]. Plant Growth

Regulation, 2007, 53(1):7-16.

[17] JIANG Y B, JI H B, LI T T, et al. Advances of the research on the phytoremediation of the selenium polluted environment[J]. Bulletin of Mineralogy, Petrology and Geochemistry, 2007, 26(1):98-104.

[18] JIANG Y, PENG Y, LI Z H, et al. Effects of paclobutrazol, uniconazole and chlorcholinchlorid on dwarfing of *Zamioculcasz amiifolia* [J]. Acta Horticulturae Sinica, 2010, 37(5):823-828.

[19] KAHIA J, SALLAH P K, DIBY L, et al. A novel regeneration system for tamarillo (*Cyphomandra betacea*) via organogenesis from hypocotyl, leaf, and root explants [J]. HortScience, 2015, 50(9):1375-1378.

[20] KANDIL H, ELEIWA M M E. Effect of the growth regulator uniconazole and salt stress on growth, yield and nutrients content of *Ammi majus* L. plant[J]. Australian Journal of Basic and Applied Sciences, 2008, 2(3):1991-8178.

[21] LI P H, ZHENG S L, CAI G Z, et al. Effects of nitrogen fertilizer with uniconazole on nutrient absorption and dry matter accumulation in potato[J]. Southwest China Journal of Agricultural Sciences, 2012, 25(6):2138-2142.

[22] LI T L, LI Y Q. Effects of calcium on activities of protective enzymes and soluble protein content in the leaves of tomato under low light stress[J]. Acta Horticulturae Sinica, 2008, 35(11):1601-1606.

[23] LI Z, FAN R, PENG X, et al. Salicylic acid alleviates selenium stress and promotes selenium uptake of grapevine [J]. Physiology and Molecular Biology of Plants, 2022, 28:625-635.

[24] LIN L J, SUN J, CUI T H, et al. Selenium accumulation characteristics of *Cyphomandra betacea* (*Solanum betaceum*) seedlings[J]. Physiology and Molecular Biology of Plants, 2020a, 26(7):1375-1383.

[25] LIN L J, WU C F, JIANG W, et al. Grafting increases cadmium accumulation in the post-grafting generations of the potential cadmium-hyperaccumulator *Solanum photeinocarpum*[J]. Chemistry and Ecology, 2020b, 36(7):685-704.

[26] LIU L Q, ZHANG Y Q, LI X, et al. Effects of seed soaking with different kinds of uniconazole on seed germination and seedling growth of adzuki bean[J]. Jiangsu Agricultural Sciences, 2017, 45(3):64-70.

[27] LIU Y, YANG F, HUANG M J, et al. Uniconazole-induced starch accumulation in

the bioenergy crop duckweed (*Landoltia punctata*) Ⅱ: transcriptome analysis of the effects of uniconazole on chlorophyll and endogenous hormone biosynthesis [J]. Biotechnology for Biofuels, 2015, 8(1): 57.

[28] LU Y, LI J, SUN J, et al. Research progress on the biological effects of uniconazole [J]. Barley and Cereal Sciences, 2017, 34(1): 6-10.

[29] MA Z, WANG X J, SUN Z Q, et al. Suitable nitrogen fertilizer rate for foliar spray of uniconazole in sweet potato [J]. Journal of Plant Nutrition and Fertilizer, 2016, 22(5): 1433-1440.

[30] MENG N, WEI S H. Uniconazole spraying ameliorates salt injury to soybean seedlings by regulating anatomical structure in roots [J]. Chinese Journal of Ecology, 2018, 37(12): 3605-3609.

[31] PEI H R, LI W, ZHANG L, et al. Research and application of plant growth regulators [J]. Shandong Agricultural Sciences, 2015, 47(7): 142-146.

[32] RASTMANESH F, MOORE F, KESHAVARZI B. Speciation and phytoavailability of heavy metals in contaminated soils in Sarcheshmeh area, Kerman Province, Iran [J]. Bulletin of Environmental Contamination and Toxicology, 2010, 85(5): 515-519.

[33] SORS T G, ELLIS D R, SALT D E. Selenium uptake, translocation, assimilation and metabolic fate in plants [J]. Photosynthesis Research, 2005, 86(3): 373-389

[34] VAN HOEWYK D. A tale of two toxicities: malformed selenoproteins and oxidative stress both contribute to selenium stress in plants [J]. Annals of Botany, 2013, 112(6): 965-972.

[35] WAN Y, YAN Y H, YANG W Y. Effects of foliar spraying uniconazole on growth and nitrogen metabolism of relay strip intercropping soybean under different nitrogen levels [J]. Journal of Zhejiang University (Agriculture and Life Sciences), 2012, 38(2): 185-196.

[36] WANG C, ZHAO H D, FENG N J, et al. Effects of S_{3307} and DTA-6 on the photosynthetic characteristics and yield of kidney bean plants in the reproductive stage [J]. Acta Prataculturae Sinica, 2018, 27(11): 162-170.

[37] WANG D D, HUANG Y, ZHOU Z Z, et al. Effects of selenate at different concentrations on growth and physiological indexes of tea tree [J]. Guihaia, 2021, 41(2): 183-194.

[38] WEI Y Z, DONG C, WANG Y, et al. Effects of uniconazole on the endogenous phytohormone content in the florescence period of Lithi [J]. Chinese Journal of Tropical Crops, 2017, 38(7): 1188-1192.

[39] WINKEL L H, JOHNSON C A, LENZ M, et al. Environmental selenium research: from microscopic processes to global understanding [J]. Environmental Science and Technology, 2012, 46(2):571-579.

[40] WU C H, NIU Y S, TANG D C, et al. Study on the accumulation law of NPK in bulbs of oriental lily under different nutritional formulas [J]. Journal of Anhui Agricultural Sciences, 2009, 37(11):4966-4967, 4976.

[41] XIE B T, ZHANG H Y, WANG B Q, et al. Effects of plant growth regulators on assimilation substance and fertilizer distribution of sweet potato under high nitrogen and water conditions [J]. Shandong Agricultural Sciences, 2020, 52(3):60-65.

[42] XU S J, SUN Z P, WANG Z X. Effects of uniconazole treatments on photosynthetic and physiological indices of aeroponic potato [J]. Acta Agriculturae Boreali-Occidentalis Sinica, 2009, (3):127-130.

[43] YAN Y H, YANG W Y, ZHANG X Q. Improve soybean seedling growth by uniconazole under shading by corn in relay strip intercropping system [J]. Chinese Journal of Oil Crop Sciences, 2011, 33(3):259-264.

[44] YIN H, QI Z, LI M, et al. Selenium forms and methods of application differentially modulate plant growth, photosynthesis, stress tolerance, selenium content and speciation in *Oryza sativa* L [J]. Ecotoxicology and Environmental Safety, 2019, 169:911-917.

[45] YUAN J R. Effects of plant growth regulators on yield, quality, N, P and K nutrition in leafy-vegetable crop [D]. University of Chinese Academy of Sciences, Beijing, China. 2013.

[46] ZHANG C Y, JIN X J, ZHANG M C, et al. Effects of S3307 and CCC mixed trace element dressing on the photosynthetic physiology and yield of soybean [J]. Soybean Science, 2020, 39(4):587-594.

[47] ZHANG M C, DUAN L S, TIAN X L, et al. Uniconazole-induced tolerance of soybean to water deficit stress in relation to changes in photosynthesis, hormones and antioxidant system [J]. Journal of Plant Physiology, 2008, 164(6):709-717.

[48] ZHONG R C, TANG X M, JIANG J, et al. Effects of uniconazole on growth, photosynthesis and yield traits of peanut [J]. Guangdong Agricultural Sciences, 2015, (11):65-70.

[49] ZHU Y, HUANG W, LIU Z, et al. Application of supramolecular nano-material adsorbent in the treatment of heavy metal pollution in acid selenium-rich soils in South China [J]. Integrated Ferroelectrics, 2021, 217(1):69-81.

Section Ⅲ Gibberellic acid

1 Introduction

Soil selenium (Se)is unevenly distributed worldwide, with the most areas lacking Se (Li et al., 2020). Se is one of the essential trace elements of human body, and the Se deficiency is threatening the human health. Eating the Se-enriched vegetables and fruits can safely and effectively increase the Se content in human body (Jiang et al., 2016). Therefore, the increase of Se content in vegetables and fruits can prevent the Se deficiency of human body (Zhao, 2019; Li et al., 2020). Most crops are non-Se accumulators with the Se content not exceeding 30 mg/kg, which cannot meet the dietary requirements for human body (Zhang et al., 2002). Se content in plants positively correlates with the exogenous Se concentration application (Jin et al., 2003; Wang et al., 2016). Various methods, such as the Se application in soil, Se seed dressing, foliar application of Se on leaves, and Se-containing nutrient solution culture, are mainly used to increase the Se content in crops (Hawkesford et al., 2007; Boldrin et al., 2013). However, these methods have many limitations, such as high cost and environmental pollution (Wu et al., 2015; Chen et al., 2018). Therefore, other methods that can improve the Se uptake ability in crops should be identified.

Se uptake in crops has been improved using agronomic measures, such as intercropping, grafting, and application of plant hormones (Luo, 2019; Huan et al., 2021; Pan et al., 2021). Plant hormones mediate the plant growth and nutrient (including Se) absorption (Peleg et al., 2011). For instance, gibberellic acid (GA) promotes the photosynthesis, vegetative growth, and fruit development of crops (Castrocamba et al., 2022). GA can regulate the crop growth and promote the nutrient absorption (Hassan et al., 2017; Zhang et al., 2020). GA induces various responses in plants under stress conditions to improve stress tolerance, thus promoting the growth of plants (Colebrook et al., 2014; Niharika et al., 2020). Moreover, GA can promote the detoxification function of antioxidant enzymes and antioxidant activity of plants to scavenge reactive oxygen species (ROS)under heavy metal stress, thus resisting the adverse reactions of heavy metals (Abolghassem et al., 2020). GA can resist the cadmium (Cd)stress by regulating the expression of various genes related to shoot and root growth, metabolism, photosynthetic pigment, and shoot and root morphology of plants (Rashid et al., 2021). GA can also promote the Cd uptake in accumulator plant *Stellaria media* and improve its phytoremediation capability (Xie et al., 2016). However, GA decreases the Cd uptake in other common plants, thus promoting their growth (Yang et al., 2021). GA promotes the nitrogen uptake and utilization rate in lettuce, thus improving the quality and yield of lettuce (Miceli et al., 2019). Furthermore, GA synthesis-related genes and proteins promotes the absorption and utilization of phosphorus in barley GA-deficient mutants (Gualano et al., 2021). GA can also promote calcium absorption after binding to certain proteins, possibly because the protein-GA complex activates Ca^{2+} channels or promotes the degradation of Ca^{2+} channel repressors through the ubiquitin-proteasome pathway. An unknown auxiliary GA receptor may also be involved in the GA-induced increase in calcium (Takeshi et al., 2018). Therefore, GA can promote the growth of plants and improve the plant stress resistance, thus changing nutrient uptake. However, a few studies have reported on the effect of GA on the Se uptake in plants under Se stress.

Cyphomandra betacea is a fruit tree belonging to the Solanaceae family, with the high ornamental and edible value (Kouame et al., 2015). A previous study showed that the Se accumulation capacity is lower in *C. betacea* than in other plants (Lin et al., 2020a). Therefore, GA application may increase the Se accumulation capacity of *C. betacea,* thus improving its commercial value. In this study, the effects of different concentrations of GA on the growth and Se uptake of *C. betacea* were studied under Se stress. This study aimed to determine the best GA concentration that could alleviate the Se stress to *C. betacea*, and promote the growth and Se accumulation of *C. betacea*. This study may provide a reference for producing Se-enriched *C. betacea*.

2 Materials and methods

2.1 Materials collections and treatments

In September 2021, the mature seeds of *C. betacea* were collected from a 5-year-old fruiting tree at the Chengdu Campus of Sichuan Agricultural University (30°42′N, 103°51′E). The seeds were air-dried, then sown in 50-hole plug seedling trays (53 cm length × 28 cm width × 10 cm height) containing moist perlite. The seedling trays were placed in the greenhouse at 25 ℃, relative humidity of 70%, and 10,000 Lux for 14 h during the day; and at 20 ℃, relative humidity of 90%, and 0 Lux for 10 h during the night (Liu et al., 2021). The perlite was irrigated with 1/2 Hoagland nutrient solution every three days after seed germination. The seedlings were transplanted into the pot after growing to about 15 cm (one month later).

GA (GA_3) was obtained from Beijing XMJ Scientific Co., Ltd. (Beijing, China).

Se used in this experiment is the analytical pure sodium selenite (Na_2SeO_3), and obtained from Nanjing Chemical Reagent Co., Ltd. (Nanjing, China).

2.2 Experimental design

Two uniform *C. betacea* seedlings were transplanted into each plastic pot (10 cm height ×15 cm diameter) filled with perlite in October 2021. The pots were also placed in the greenhouse under the conditions described above. Hoagland nutrient solution (100 mL) was irrigated into each pot after transplantation for 7 days. GA solution (0, 100, 200, 300, and 400 mg/L) (Yang et al., 2021) was fully sprayed on both sides of the leaves (10 mL per pot). The spraying was conducted again after 15 days. Each treatment had three replicates (three pot as one repetition, and 45 pots in total). Meanwhile, each pot was irrigated with Hoagland nutrient solution (100 mL) containing 0.1 mg/L Se in the form of Na_2SeO_3 (Liu et al., 2019) every 3 days during the whole plant growth. The positions of pots were randomly changed to reduce the influence of marginal effects during the whole growth process. The plants were harvested after 30 days of the first GA treatment to determine the various indicators.

2.3 Determination of indicators

In November 2021, the fifth mature leaves (from the top) of each plant were chosen to measure the net photosynthetic rate (Pn), transpiration rate (Tr), stomatal conductance (Gs), and intercellular CO_2 concentration (Ci) using a Li-6400 photosynthetic system (LI-COR, Lincoln, NE, USA). The Li-6400 photosynthetic system and measurement time were conducted as in the previous report (Li et al., 2022). The same leaves were used to determine the content of photosynthetic pigments (chlorophyll a, chlorophyll b, and carotenoid) and antioxidant enzymes [superoxide dismutase (SOD), peroxidase (POD), and catalase (CAT)] activities. The photosynthetic pigment content was determined using the ethanol and acetone extraction method as in the previous report (Lin et al., 2020b). Pretreatments were conducted as in the previous report (Lin et al., 2020b) for the analysis of antioxidant enzyme activity. The activities of SOD, POD, and CAT, were determined using the nitroblue tetrazole, guaiacol colorimetric, and potassium permanganate titration methods, respectively (Hao et al., 2004). The plants were harvested, washed, and then dried as in the previous report (Li et al., 2022). The root and shoot biomass (dry weight) were measured using an electronic balance with an accuracy of one thousandth. The dry plant samples were finely ground, soaked with the nitrate acid and perchloric acid for 12 h, digested until transparent at a hot plate, and then reduced by the hydrochloric acid. The digested solution was made up to 50 mL using the deionized water, then used to determine the Se content via a hydride generation-atomic fluorescence spectrometry (AFS-9700, Beijing Haiguang Instrument Co., Ltd., Beijing, China) (Li et al., 2022). The root/shoot biomass (dry weight) ratio and chlorophyll a/b ratio were calculated. The translocation factor (TF) was calculated as follows: TF = Se content in shoots/Se content in roots (Rastmanesh et al., 2010).

2.4 Statistical analyses

The software SPSS 20.0 (IBM, Chicago, IL, USA) was used for all statistical analyses. All data with three biological repetitions was subjected to a normal distribution and homogeneity tests before one-way analysis of variance followed by the Duncan's Multiple Range Test ($P < 0.05$). The relationship between GA concentration and biomass and Se content was analyzed using regression analysis. Pearson's correlation was used to analyze the relationships among all items. The grey relational analysis was performed according to the previous reports (Wang, 2019a; Ma et al., 2022).

3 Results

3.1 Biomass

Compared with the control, GA increased the root and shoot biomass of *C. betacea* (Figure 3-6 and Figure 3-7). The root and shoot biomass increased with the increase of GA concentrations up to 300 mg/L, and then decreased when the GA concentration was higher than 300 mg/L. The root biomass and shoot biomass were got the maximums when the GA concentration was 300 mg/L. Compared with the control, the concentrations of 100, 200, 300, and 400 mg/L GA increased the root biomass by 23.69%, 29.65%, 54.29%, and 33.21%, respectively, and increased the shoot biomass by 29. 51%, 33. 29%, 59. 51%, and 40. 46%, respectively. Additionally, the root biomass had a quadratic polynomial regression relationship with the GA concentration ($y=-2.864E^{-7}x^2+0.056$, $R^2=0.814$, $P=0.000$), and the shoot biomass also had a quadratic polynomial regression relationship with the GA concentration ($y=-1.729E^{-6}x^2+0.001x+0.323$, $R^2=0.860$, $P=0.001$). However, GA did not significantly affect the root/shoot ratio of *C. betacea* (Figure 3-8).

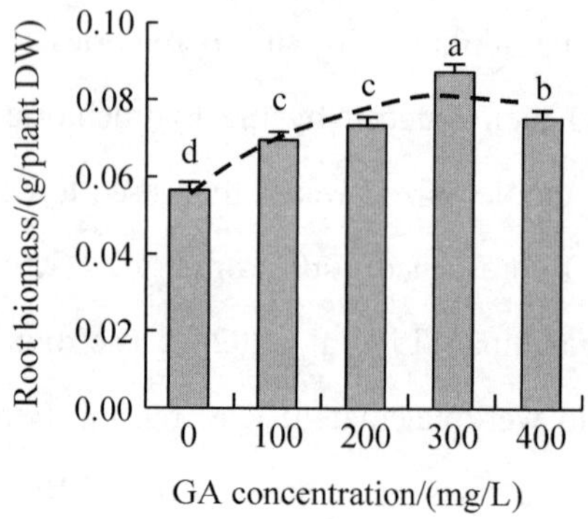

Note: Different lowercase letters indicate significant differences among the treatments (Duncan's multiple range test, $P < 0.05$).

Figure 3-6 Root biomass of *C. betacea*

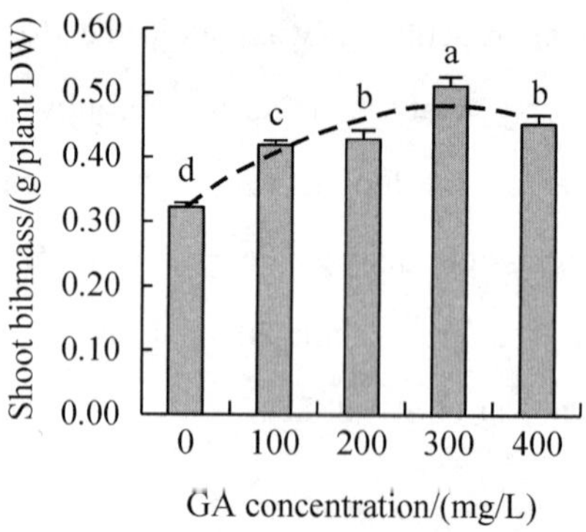

Note: Different lowercase letters indicate significant differences among the treatments (Duncan's multiple range test, $P < 0.05$).

Figure 3-7 Shoot biomass of *C. betacea*

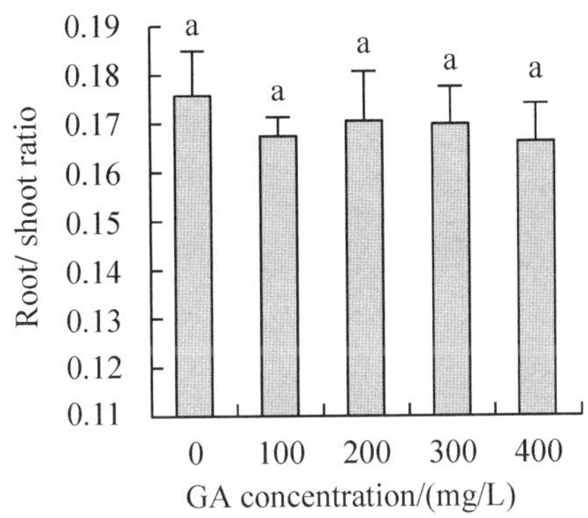

Note: Different lowercase letters indicate significant differences among the treatments (Duncan's multiple range test, $P < 0.05$).

Figure 3-8　Root/shoot ratio of *C. betacea*

3.2　Photosynthetic pigment content

Compared with the control, GA increased the chlorophyll a, chlorophyll b, and carotenoid content in *C. betacea* to some extent (Figure 3-9, Figure 3-10, and Figure 3-11). The content of chlorophyll a, chlorophyll b, and carotenoid increased with increasing GA concentration up to 300 mg/L, and then decreased. When the GA concentration was 300 mg/L, the content of chlorophyll a, chlorophyll b, and carotenoid got the maximums, which increased by 36.91%, 42.96%, and 116.91%, respectively, compared with the control.

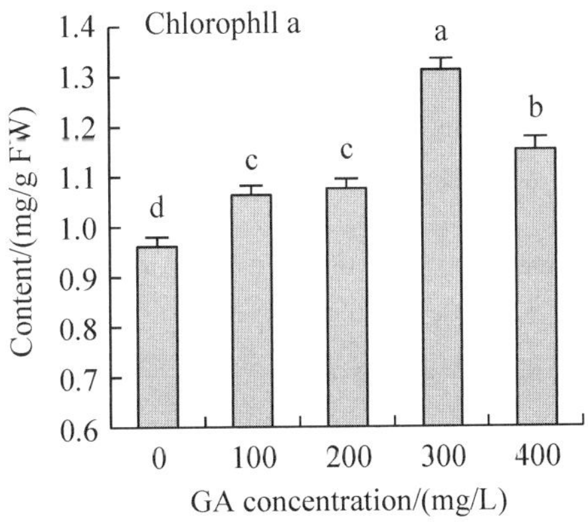

Note: Different lowercase letters indicate significant differences among the treatments (Duncan's multiple range test, $P < 0.05$).

Figure 3-9　Chlorophyll a content in *C. betacea*

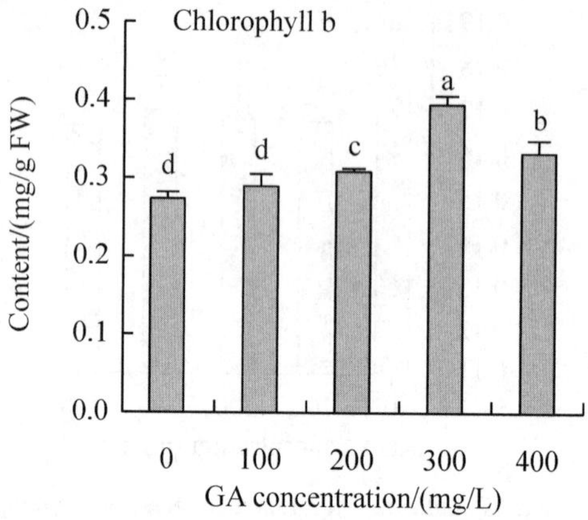

Note: Different lowercase letters indicate significant differences among the treatments (Duncan's multiple range test, $P < 0.05$).

Figure 3-10 Chlorophyll b content in *C. betacea*

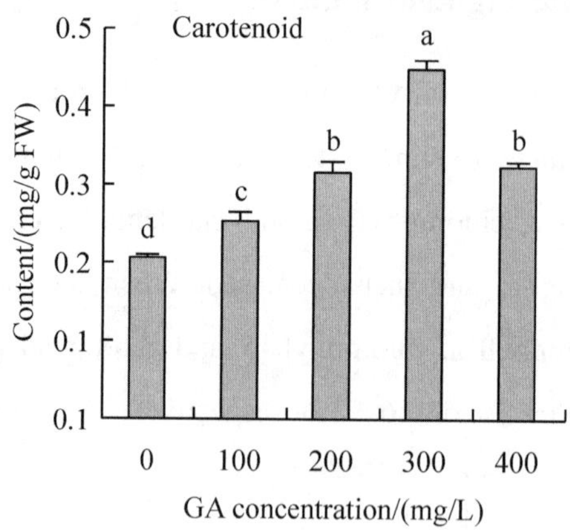

Note: Different lowercase letters indicate significant differences among the treatments (Duncan's multiple range test, $P < 0.05$).

Figure 3-11 Carotenoid content in *C. betacea*

3.3 Photosynthetic characteristics

With the increase of GA concentration, the Pn, Gs, Ci, and Tr of *C. betacea* increased when the GA concentration was not higher than 300 mg/L, and then decreased when the GA concentration was higher than 300 mg/L (Table 3-10). GA increased the Pn, Gs, Ci, and Tr to some extent, compared with the control. When the GA concentration was 300 mg/L, the Pn, Gs, Ci, and Tr got the maximums, which increased by 82.81%, 81.01%, 30.35%, and 72.44%, respectively, compared with the control.

Table 3-10 Photosynthetic characteristic of *C. betacea*

GA concentration/(mg/L)	Pn/[μmol CO_2/(m^2/s)]	Gs/[mol H_2O/(m^2/s)]	Ci/(μmol CO_2/mol)	Tr/[mmol H_2O/(m^2/s)]
0	4. 120±0. 097e	0. 0437±0. 0008c	194. 4±4. 26d	0. 976±0. 015e
100	5. 209±0. 104d	0. 0448±0. 0019c	209. 1±8. 25c	1. 021±0. 011d
200	6. 482±0. 165c	0. 0631±0. 0033b	216. 1±6. 11c	1. 362±0. 020c
300	7. 532±0. 117a	0. 0791±0. 0022a	253. 4±9. 47a	1. 683±0. 014a
400	6. 835±0. 059b	0. 0765±0. 0016a	232. 5±6. 49b	1. 626±0. 021b

Note: Different lowercase letters indicate significant differences among the treatments (Duncan's multiple range test, $P < 0.05$). Pn = net photosynthetic rate; Tr = transpiration rate; Gs = stomatal conductance; Ci = intercellular CO_2 concentration.

3.4 Antioxidant enzyme activity

Compared with the control, GA increased the SOD, CAT, and POD activities of *C. betacea* to some extent (Figure 3-12, Figure 3-13, and Figure 3-14). As same as the photosynthetic characteristics, the activities of SOD, CAT, and POD had the increase trend with the increase of GA concentration when the GA concentrations were not higher than 300 mg/L, and then had the decrease trend when the GA concentration was higher than 300 mg/L. The maximums of SOD, CAT, and POD activities were at the GA concentration of 300 mg/L, which increased by 107. 03%, 18. 20%, and 40. 37%, respectively, compared with the control.

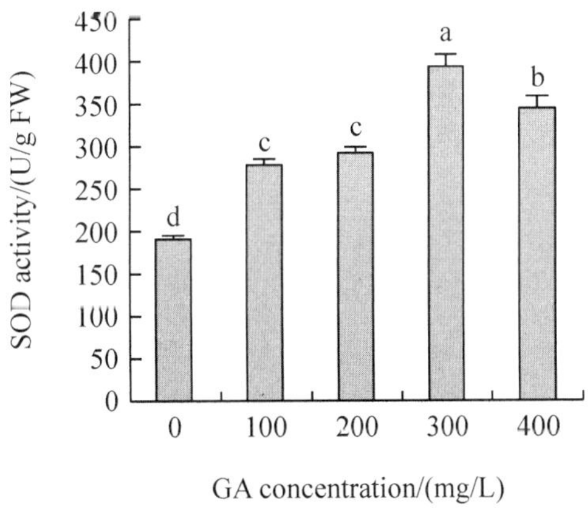

Note: Different lowercase letters indicate significant differences among the treatments (Duncan's multiple range test, $P < 0.05$).

Figure 3-12 SOD activity of *C. betacea*

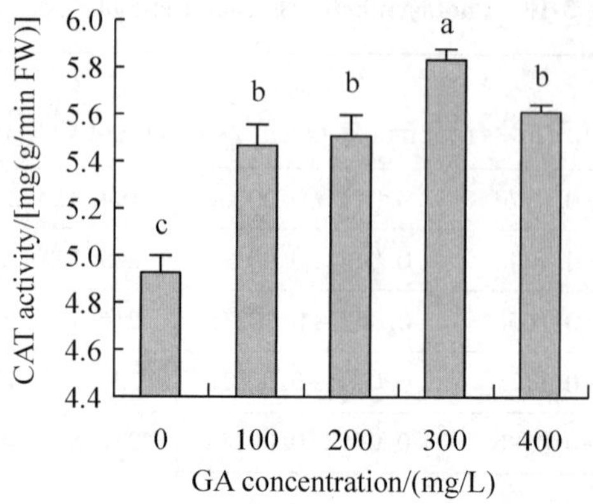

Note: Different lowercase letters indicate significant differences among the treatments (Duncan's multiple range test, $P < 0.05$).

Figure 3-13 CAT activity of *C. betacea*

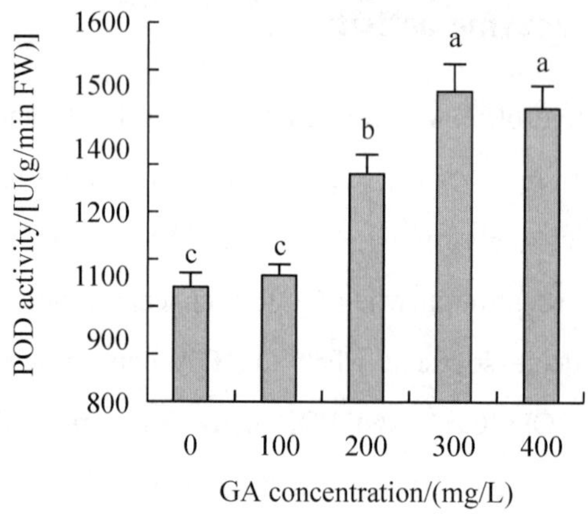

Note: Different lowercase letters indicate significant differences among the treatments (Duncan's multiple range test, $P < 0.05$).

Figure 3-14 POD activity of *C. betacea*

3.5 Se content

Although GA increased the Se content in roots and shoots of *C. betacea*, the Se content as higher in roots than in shoots at the different GA concentrations (Figure 3-15 and Figure 3-16). The Se content were increased in roots and shoots with increasing GA concentrations up to 300 mg/L, then decreased. The maximums of Se content were at the GA concentration of 300 mg/L. Compared with the control, the GA concentrations of 100, 200, 300, and 400 mg/L increased the root Se content by 9.42%, 38.95%, 70.31%, and 29.39%, respectively, and the shoot Se content by 4.68%, 8.83%, 22.02%, and 11.85%, respectively. Furthermore, the root content had a quadratic polynomial regression relationship with the GA concentration ($y = -7.391E^{-5}x^2 + 0.042x + 9.583$, $R^2 = 0.696$, $P = 0.001$), and the shoot Se content also had a quadratic polynomial regression relationship with the GA concentration ($y = -1.655E^{-6}x^2 + 0.001x + 1.105$, $R^2 = 0.683$, $P = 0.001$). GA decreased the TF of *C. betacea*, and the TF had an opposite trend to the Se content (Figure 3-17).

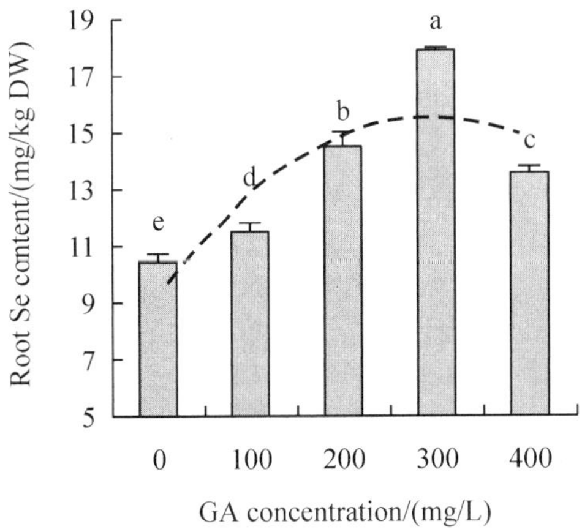

Note: Different lowercase letters indicate significant differences among the treatments (Duncan's multiple range test, $P < 0.05$).

Figure 3-15 Root Se content in *C. betacea*

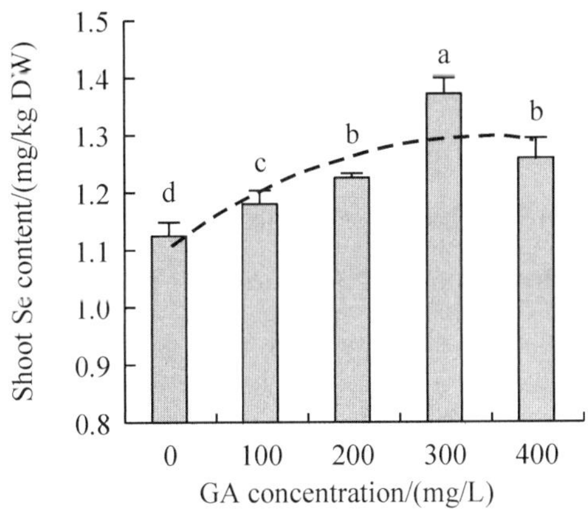

Note: Different lowercase letters indicate significant differences among the treatments (Duncan's multiple range test, $P < 0.05$).

Figure 3-16 Shoot Se content in *C. betacea*

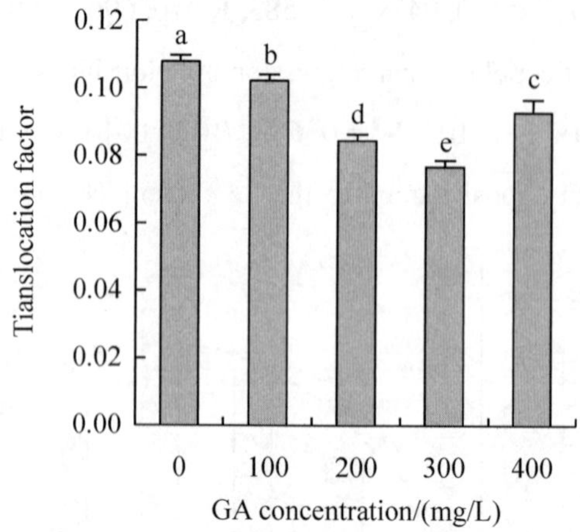

Note: Different lowercase letters indicate significant differences among the treatments (Duncan's multiple range test, $P < 0.05$).

Figure 3-17 Translocation factor of *C. betacea*

3.6 Correlation and grey relational analyses

Correlation analysis showed that the biomass, Se content, photosynthetic pigment content, photosynthetic characteristic parameters, and antioxidant enzyme activities were significantly positively correlated with each other (Table 3-11). Grey relational analysis showed that the biomass, photosynthetic pigment content, photosynthetic characteristic parameters, and antioxidant enzyme activities, and root Se content were correlated with the shoot Se content (Figure 3-18). The order of grey correlation coefficient was as follows: carotenoid content > chlorophyll a content > chlorophyll b content > root Se content > SOD activity > POD activity > Tr > Pn > shoot biomass > Ci > Gs > CAT activity > root biomass.

Table 3-11 Correlations among the different indicators

Indicator	Root biomass	Shoot biomass	Chlorophyll a content	Chlorophyll b content	Carotenoid content	Pn	Gs	Ci	Tr	SOD activity	CAT activity	POD activity	Root Se content	Shoot Se content
Root biomass														
Shoot biomass	0.954 * *													
Chlorophyll a content	0.936 * *	0.931 * *												
Chlorophyll b content	0.881 * *	0.857 * *	0.955 * *											
Carotenoid content	0.923 * *	0.927 * *	0.961 * *	0.960 * *										
Pn	0.933 * *	0.940 * *	0.898 * *	0.869 * *	0.923 * *									
Gs	0.817 * *	0.832 * *	0.842 * *	0.862 * *	0.870 * *	0.942 * *								
Ci	0.893 * *	0.899 * *	0.924 * *	0.945 * *	0.933 * *	0.894 * *	0.892 * *							
Tr	0.840 * *	0.849 * *	0.857 * *	0.872 * *	0.884 * *	0.950 * *	0.994 * *	0.901 * *						
SOD activity	0.948 * *	0.975 * *	0.939 * *	0.897 * *	0.926 * *	0.949 * *	0.889 * *	0.956 * *	0.907 * *					
CAT activity	0.965 * *	0.959 * *	0.885 * *	0.792 * *	0.844 * *	0.923 * *	0.793 * *	0.847 * *	0.812 * *	0.943 * *				
POD activity	0.833 * *	0.823 * *	0.841 * *	0.876 * *	0.879 * *	0.930 * *	0.960 * *	0.875 * *	0.974 * *	0.883 * *	0.792 * *			
Root Se content	0.914 * *	0.889 * *	0.899 * *	0.919 * *	0.969 * *	0.913 * *	0.848 * *	0.883 * *	0.860 * *	0.872 * *	0.830 * *	0.852 * *		
Shoot Se content	0.907 * *	0.924 * *	0.953 * *	0.945 * *	0.954 * *	0.911 * *	0.869 * *	0.899 * *	0.880 * *	0.914 * *	0.866 * *	0.872 * *	0.938 * *	

Note: * * = Correlation is significant at the 0.01 level (2-tailed test). N = 15. Pn = net photosynthetic rate; Gs = stomatal conductance; Ci = intercellular CO_2 concentration; Tr = transpiration rate.

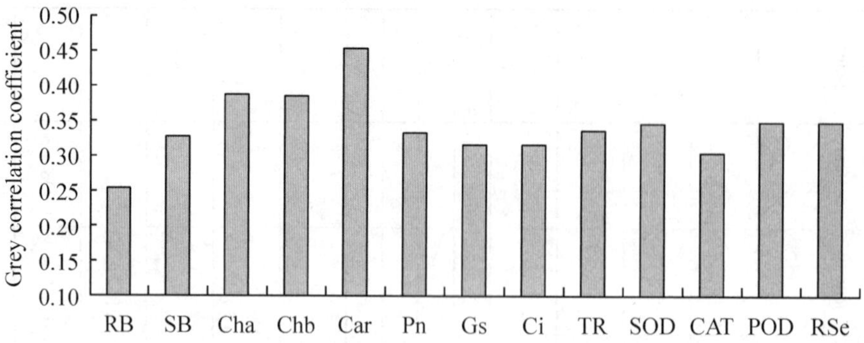

Note: RB = root biomass; SB = shoot biomass; Cha = chlorophyll a content; Chb = chlorophyll b content; Car = carotenoid content; Pn = net photosynthetic rate; Gs = stomatal conductance; Ci = intercellular CO_2 concentration; Tr = transpiration rate; SOD = SOD activity; POD = POD activity; CAT = CAT activity; RSe = root Se content.

Figure 3-18 Grey correlation coefficients of different parameters with the shoot Se content

4 Discussion

GA participates in various plant life activities and is closely related to the metabolic regulation of plant growth and development (Gao et al., 2018). Previous studies have shown that appropriate concentrations of GA could increase the growth of stem length, stem thickness, and leaf area of plum, tomato, and sweet pepper seedlings (Tahani et al., 2019; Miceli et al., 2020). In this experiment, GA increased the root and shoot biomass of *C. betacea* under Se stress, which is consistent with the previous studies (Tahani et al., 2019; Miceli et al., 2020;). Additionally, the biomass of root and shoot had a quadratic polynomial regression relationship with the GA concentration. This could be due to: First, GA directly regulates the content of aspartic acid-glutamic acid-leucine-leucine-alanine (DELLA) protein in the plant nucleus, thus promoting the plant growth via ubiquitination and degradation of DELLA protein by binding to the specific receptor gibberellin insensitive dwarf 1 (GID1). DELLA protein is a key regulator in the GA signaling pathway, inhibiting the expression of downstream genes, thus inhibiting the elongation of plant hypocotyls and stems (Bai et al., 2019; Jorge et al., 2021). The exogenous application of GA increases the concentration of extracellular active GA in plants, thus enhancing the degradation of DELLA protein (Bai et al., 2019). Second, GA interacts with other plant hormones, thus indirectly affecting the plant growth by changing the levels of auxin, brassinolide, and ethylene (Wang et al., 2017; Castrocamba et al., 2022).

Chlorophyll and carotenoid participate in photosynthesis by capturing light energy, driving electron transfer to reaction centers, and defending against the photo damage (Petra et al., 2003). Furthermore, the photosynthetic carbon assimilation ability of plants is closely related to the photosynthetic traits. Therefore, the photosynthetic indicators can reflect the growth status and productivity of plants (Sun et al., 2017). Under Se stress, the content of chlorophyll a, chlorophyll b, and carotenoid in *C. betacea* decreases, and has a decrease trend with the increase of Se concentration (Lin et al., 2020a). Studies have shown that a high concentration (higher than 40 mg/L) of GA decreases the chlorophyll a and carotenoid content, Pn, Ci, Gs, and Tr in spinach seedling leaves, while a low concentration (lower than 40 mg/L)of GA had the opposite effect under copper stress (Gong et al., 2021). Under Cd stress, GA also increases the photosynthetic pigment content and photosynthetic parameters of *C. betacea* (Yang et al., 2021). In this study, the content of chlorophyll a, chlorophyll b, and carotenoid in leaves of *C. betacea* increased with increasing GA concentration up to 300 mg/L, and then decreased under Se stress. Moreover, the changes in Pn, Ci, Gs, and Tr of *C. betacea* were similar to that of the photosynthetic pigment content. These results are consistent with the previous studies (Gong et al., 2021; Yang et al., 2021), indicating that the GA could promote the synthesis of photosynthetic pigments and improve the photosynthesis of *C. betacea* under Se stress. Because the GA can promote the transduction of GA signal in plants, up-regulates the expression of photosynthetic pigment synthesis-related genes, and down-regulates the expression of photosynthetic pigment metabolism-related genes, thereby increases the photosynthetic pigment content (Keawmanee et al., 2022).

Under stress condition, the excessive ROS is produced in plants, which may damage the cellular components, disrupt cell membranes, and inhibit the growth of plants (Labanowska et al., 2012). Se is similar to toxic metals, which also causes the excessive ROS to inhibit the growth of plants (Spallholz et al., 2002). Under Se stress, the POD and CAT activities of *C. betacea* increases, and have an increase trend with the increase of Se concentration (Lin et al., 2020a). Previous studies have shown that an appropriate concentration of GA increased the antioxidant enzyme activity of spinach under copper stress and okra under NaCl stress (Wang et al., 2019b; Gong et al., 2021). GA also increases the SOD, POD, and CAT activities of *C. betacea* under Cd stress (Yang et al., 2021). Herein, GA increased the SOD, POD, and CAT activities in the leaves of *C. betacea* under Se stress, which is consistent with the previous studies (Wang et al., 2019b; Gong et al., 2021; Yang et al., 2021). These results indicate that GA could improve the resistance of *C. betacea* under Se stress, which may be related to the expressions of antioxidative enzyme-related genes regulating by GA (Peng et al., 2022).

Plants mainly absorb Se from soil through their roots. Soil is the main source of Se for plants (Jiang et al., 2016). Besides soil environment, the form of Se also affects the Se absorption in plants. The transport mechanism and efficiency of different valence Se in plants are different (Wang et al., 2014; Jiang et al., 2016). Some studies have assessed the characteristics of plant hormones on the absorption of Se. For instance, indole acetic acid (IAA) could promote the growth of *C. betacea* seedlings and increase the Se content and bioconcentration factor in various organs (Huan et al., 2021). Methyl jasmonate can promote the absorption of low-concentration sodium selenate in tea, thus improving the antioxidant effect (Zhou et al., 2012). Herein, GA increased the Se content in roots and shoots of *C. betacea*. In addition, the Se content in roots and shoots had a quadratic polynomial regression relationship with the GA concentration. These results indicate that GA could promote the Se absorption in *C. betacea*. Additionally, the Se content was higher in roots than that in shoots of *C. betacea*, indicating that Se is mainly concentrated in roots of *C. betacea*. This could be because sodium selenite was the exogenous Se applied. Sodium selenite mainly exists in the form of selenomethionine (organic Se) in plants, which is easily fixed in plant roots (Zhu et al., 2009). Moreover, correlation analysis showed that the biomass, Se content, photosynthetic pigment content, photosynthetic characteristic parameters, and antioxidant enzyme activities were significantly positively correlated with each other. Grey relational analysis also showed that carotenoid, chlorophyll a, and chlorophyll b content was the most related to the shoot Se content. These results show that photosynthetic pigment content was closely related to the Se uptake in *C. betacea* with the GA application. However, further studies should assess the action mechanisms.

5 Conclusion

Under Se stress, GA increases the biomass, photosynthetic pigment content, photosynthetic rate, and antioxidant enzyme activities of *C. betacea*, thereby promoting the growth of *C. betacea*. GA also increases the Se content in *C. betacea*, peaking at 300 mg/L GA. The biomass and Se content had a quadratic polynomial regression relationship with the GA concentration. Among these indicators, the carotenoid, chlorophyll a, and chlorophyll b content were the most related to the shoot Se content. The future work should assess the mechanism underlying GA promoting effects on the Se uptake in *C. betacea*.

References

[1]ABOLGHASSEM E, YULONG D, AND FARZAD M. The role of salicylic acid and gibberellin signaling in plant responses to abiotic stress with an emphasis on heavy metals [J]. Plant Signaling & Behavior, 2020, 15(7):1559-2316.

[2]BAI Y H, ZHU X D, FAN X C, et al. Research progress of plant DELLA proteins and its response to gibberellin signal regulating plant growth and development[J]. Molecular Plant Breeding, 2019, 17(8):2509-2516.

[3] BOLDRIN P F, FAQUIN V, RAMOS S J, et al. Soil and foliar application of selenium in rice biofortification[J]. Journal of Food Composition and Analysis, 2013, 31(2): 238-244.

[4] CASTRO-CAMBA R, SáNCHEZ C, VIDAL N, et al. Interactions of gibberellins with phytohormones and their role in stress responses[J]. Horticulturae, 2022, 8(241):241.

[5]CHEN Y Z, BAI Y D, JIA L G, et al. Significance of selenium-enriched vegetable research and its exploitation[J]. Northern Horticulture, 2018, (18):152-156.

[6]COLEBROOK E H, THOMAS S G, PHILLIPS A L, et al. The role of gibberellin signaling in plant responses to abiotic stress[J]. The Journal of Experimental Biology, 2014, 217(1):67-75.

[7]GAO X H, FU X D. Research progress for the gibberellin signaling and action on plant growth and development[J]. Biotechnology Bulletin, 2018, 34(7):1-13.

[8] GONG Q, LI Z H, WANG L, et al. Gibberellic acid application on biomass, oxidative stress response, and photosynthesis in spinach (*Spinacia oleracea* L.) seedlings under copper stress [J]. Environmental Science and Pollution Research, 2021, 28 (38):53594-53604.

[9] GUALANO L, MORICONI J I, OLIFERUK S, et al. Barley plants carrying the altered function Sln1d allele display modified responses to low phosphorus supply: implications for phosphorus utilization efficiency [J]. Functional Plant Biology, 2021, 48 (8):780-792.

[10]HAO Z B, CANG J, XU Z. Plant physiology experiment[M]. Harbin Institute of

Technology Press, Harbin, China, 2004.

[11] HASSAN M, MANSOOR S. Priming seeds with phytohormones alleviates cadmium toxicity in mung bean (*Vigna Radiata* L. Wilczek)seedlings[J]. Pakistan Journal of Botany, 2017, 49(6):2071-2078.

[12]HAWKESFORD M J, ZHAO F J. Strategies for increasing the selenium content of wheat[J]. Journal of Cereal Science, 2007, 46(3):282-292.

[13]HUAN Y M, YANG L, LIU Q, et al. Effects of indole acetic acid on the growth and selenium absorption characteristics of *Cyphomandra betacea* seedlings[J]. Acta Physiologiae Plantarum, 2021, 43(5):1-8.

[14]JIANG Y, ZENG Z H, YANG Q S, et al. Selenium (Se)uptake and transformation mechanisms and physiological function in plant:A review[J]. Journal of Applied Ecology, 2016, 27(12):4067-4076.

[15]JIN M Z, XING D Y, HU W H, et al. The function of selenium and exploitation and utilization of selenium resources Ⅱ:exploitation and utilization of selenium in agriculture [J]. Journal of Hubei Agricultural College, 2003, (5):392-396.

[16] JORGE H, ASIER B M, MIGUEL A B. Origin and evolution of gibberellin signaling and metabolism in plants[J]. Seminars in Cell & Developmental Biology, 2021, 109 (Special SI):46-54.

[17] KEAWMANEE N, MA G, ZHANG L C, et al. Exogenous gibberellin induced regreening through the regulation of chlorophyll and carotenoid metabolism in Valencia oranges[J]. Plant Physiology and Biochemistry, 2022, 173, 14-24.

[18]KOUAME C, KAHIA J, DIBY L, et al. A novel regeneration system for tamarillo (*Cyphomandra betacea*) via organogenesis from hypocotyl, leaf, and root explants[J]. HortScience, 2015, 50(9), 1375-1378.

[19]LABANOWSKA M, FILEK M, KOSCIELNIAK J, et al. The effects of short-term selenium stress on Polish and Finnish wheat seedlings-EPR, enzymatic and fluorescence studies[J]. Journal of Plant Physiology, 2012, 169(3):275-284.

[20][6]LI A R, QIAO X X, ZHAO F F, et al. Research progress on the relationship between selenium and human health[J]. Journal of Green Science and Technology, 2020, (12):121-122.

[21]LI Z Y, FAN R, PENG X M, et al. Salicylic acid alleviates selenium stress and promotes selenium uptake of grapevine[J]. Physiology and Molecular Biology of Plants,

2022, 28(3):625-635.

[22] LIN L J, SUN J, CUI T H, et al. Selenium accumulation characteristics of *Cyphomandra betacea* (*Solanum betaceum*)seedlings[J]. Physiology and Molecular Biology of Plants, 2020a, 26(7):1375-1383.

[23] LIN L J, WU C F, JIANG W, et al. Grafting increases cadmium accumulation in the post-grafting generations of the potential cadmium-hyperaccumulator*Solanum photeinocarpum*[J]. Chemistry and Ecology, 2020b, 36(7):685-704.

[24] LIU L, HAN J X, DENG L L, et al. Effects of diethyl aminoethyl hexanoate on the physiology and selenium absorption of grape seedlings[J]. Acta Physiologiae Plantarum, 2021, 43(8):115.

[25] LIU L, WANG T, SUI L Y, et al. The selenium accumulation characteristics of grape seedlings[J]. IOP Conference Series: Earth and Environmental Science, 2019, 330(4):042042.

[26] LUO H. Effects of mutual grafting on the growth and selenium enrichment of cherry tomato progeny of two genotypes[M]. Sichuan Agricultural University, Chengdu, China, 2019.

[27] MA Q Q, FASIH U H, MUHAMMAD F, et al. Selenium treated foliage and biochar treated soil for improved lettuce (*Lactuca sativa* L.)growth in Cd-polluted soil[J]. Journal of Cleaner Production, 2022, 335, 130267.

[28] MICELI A, MONCADA A, SABATINO L, et al. Effect of gibberellic acid on growth, yield, and quality of leaf lettuce and rocket grown in a floating system[J]. Agronomy, 2019, 9(7):382.

[29] MICELI A, VETRANO F, MONCADA A. Effects of foliar application of gibberellic acid on the salt tolerance of tomato and sweet pepper transplants[J]. Horticulturae, 2020, 6(93):93.

[30] NIHARIKA, SINGH N B, SINGH A, et al. Mitigating strategies of gibberellins in various environmental cues and their crosstalk with other hormonal pathways in plants: a review[J]. Plant Molecular Biology Reporter, 2020, 39(1):34-49.

[31] PAN S K, LU R H, LI H Y, et al. Mutual intercropping affects selenium uptake of eggplant seedlings[J]. International Journal of Environmental Analytical Chemistry, 2021, 101(15):2866-2875.

[32] PELEG Z, BLUMWALD E. Hormone balance and abiotic stress tolerance in crop

plants[J]. Current Opinion in Plant Biology, 2011, 14(3):290-295.

[33] PENG L, LUO Y, LI F, et al. Effects of exogenous methyl jasmonate and gibberellins on growth, related enzyme activities and content of active constituents of Rubia cordifolia[J]. Chinese Traditional and Herbal Drugs, 2022, 53(11):3463-3471.

[34] PETRA F, ALEXANDER M, PATRICK J, et al. Structure and function of photosystem I:interaction with its soluble electron carriers and external antenna systems[J]. FEBS Letters, 2003, 555(1):40-44.

[35] RASHID H H, SHUBHAM S, MOHAMMAD U, et al. Gibberellins target shoot-root growth, morpho-physiological and molecular pathways to induce cadmium tolerance in *Vigna radiata* L. [J]. Agronomy, 2021, 11(5):896.

[36] RASTMANESH F, MOORE F, KESHAVARZI B. Speciation and phytoavailability of heavy metals in contaminated soils in Sarcheshmeh area, Kerman Province, Iran[J]. B. Bulletin of Environmental Contamination and Toxicology, 2010, 85(5):515-519.

[37] SPALLHOLZ J E, HOFFMAN D J. Selenium toxicity:cause and effects in aquatic birds[J]. Aquatic Toxicology, 2002, 57(1):27-37.

[38] SUN M, TIAN K, ZHANG Y, et al. Research on leaf functional traits and their environmental adaptation[J]. Plant Science Journal, 2017, 35(6):940-949.

[39] TAHANI J M A, MANAR I A, ZAINAB O O. Effect of spraying of gibberellic acid (GA_3) and foliar fertilizer (Agroleaf) on the seedling growth of plum cultivar "Hollywood"[J]. Research on Crops, 2019, 20(2):313-321.

[40] TAKESHI I, KANAKO O, JUTAROU F, et al. DELLA-dependent and -independent gibberellin signaling[J]. Plant Signaling & Behavior, 2018, 13(3):e1445933.

[41] WANG Q Y, ZHANG W, WANG H P, et al. Selnium, lead, cadmium and zinc uptake in soils with different content of selenium of three types of vegetables[J]. Chinese Journal of Tropical Crops, 2016, 37(4):672-678.

[42] WANG X F, CHEN S Y, LUO Z, et al. Mechanisms of selenium uptake, translocation and chemical speciation transformation in plants[J]. Journal of Agricultural Resources and Environment, 2014, 31(6):539-544.

[43] WANG X L. Application of grey relation analysis theory to choose high reliability of the network node[J]. Journal of Physics:Conference Series, 2019a, 1237(3):032056.

[44] WANG Y H, ZHANG G, CHEN Y, et al. Exogenous application of gibberellic acid and ascorbic acid improved tolerance of okra seedlings to NaCl stress[J]. Acta Physiologiae

Plantarum, 2019b, 41(6):93.

[45] WANG Y J, ZHAO J, LU W J, et al. Gibberellin in plant height control: old player, new story[J]. Plant Cell Reports, 2017, 36(3):391-398.

[46] WU Z L, BANUELOS G S, LIN Z Q, et al. Biofortification and phytoremediation of selenium in China[J]. Frontiers in Plant Science, 2015, 6(3):136.

[47] XIE Y D, LIU C Y, YU X N, et al. Effects of exogenous GA_3 on growth and cadmium accumulation of accumulator plant *Stellaria media*[J]. Chinese Journal of Soil Science, 2016, 47(6):1363- 1367.

[48] YANG L, HUAN Y M, SUN J, et al. Effects of exogenous gibberellic acid on growth and cadmium accumulation in *Cyphomandra betacea* seedlings[J]. Environmental Progress & Sustainable Energy, 2021, 40(5):e13655.

[49] ZHANG C Y, WANG H J, WANG H B. Mechanisms of plant tolerance to heavy metals mediated by gibberellic acid[J]. Journal of Ecology and Rural Environment, 2020, 36(2):137-144.

[50] ZHANG C, WU Y Y, PENG Z K. Progress in the research of selenium in plants[J]. Journal of Hubei Minzu University (Nature Seience Ediion), 2002, 20, 58-62.

[51] ZHAO L. Nutritional value of selenium and present situation of development and utilization of selenium-rich agricultural products[J]. Science and Technology of Qinghai Agriculture and Forestry, 2019, (1):50-53.

[52] ZHOU W, WEI C X, WEI W. Influence of methyl jasmonate on Se content and quality of selenium-enriched spring tea[J]. Journal of Mountain Agriculture and Biology, 2012, (6):514-518.

[53] ZHU Y G, PILON-SMITS E A H, ZHAO F J, et al. Selenium in higher plants: understanding mechanisms for biofortification and phytoremediation[J]. Trends in Plant Science, 2009, 14(8):436-442.

Section Ⅳ Indole acetic acid

1 Introduction

Selenium (Se), a component of glutathione peroxidase (GSH-Px), has been proven to be a necessary micronutrient for the human body (Yuan et al., 2016; Wang, 2010), and studies have shown that an exceedingly low level of Se in the body may be one of the causes of some major diseases like Keshan disease (Liu et al., 2014; Zhang et al., 2018). The human body can ingest Se through food, and soil is the most important source of Se (Li et al., 2016). Plants have become important carriers of dietary Se supplements because of their ability to absorb and convert inorganic Se from the soil into organic Se, which is effective and safe for the human body. Therefore, a research focus on adopting technologies and measures to improve the absorption of Se by plants is merited.

Indole acetic acid (IAA), also known as auxin, was the earliest hormone discovered by humans to promote plant growth (Zou et al., 2018). It is primarily produced in young buds, the leaves of plants and developing seeds. Although IAA is a small molecule with a simple structure, and it is involved in the growth and development of different organs of plants (Woodward et al., 2005; Korasick et al., 2013). It is determined that IAA can not only promote seed germination (Liu et al., 2018) but also improve photosynthesis (Dong et al., 2001) and increase the biomass by plants (Kai et al., 2007). Moreover, IAA also affects the absorption of some elements by plants. Under nutrition stress of low nitrogen or potassium, IAA can enhance the nitrogen, phosphorus (Ning, 2013) and potassium (Guo et al., 2019) uptake by plants. The application of exogenous IAA can improve antioxidant enzyme activity (Liu et al., 2007) and alleviate the toxic effect of heavy metals on plants, enhancing the tolerance and regulating the absorption of lead, zinc, cadmium, manganese, and nickel (Bashri et al., 2016; Bulak et al., 2014; Fassler et al., 2010; Liu et al., 2007). It can be observed that IAA has been increasingly applied to promote the growth of plants and improve the plant adaptability to the environment, which has resulted in corresponding progress and achievement. However, there are few studies on the improvement of the ability of plants to absorb Se in concert with exogenous IAA.

As a fruit tree, *Cyphomandra betacea* grows naturally in areas of cool climate, which is strongly adaptable, easily cultivated, and possesses a long fruiting period and storage resistance (Hassan et al., 2013). In addition, *C. betacea* is suitable for processing into various foods whose fruits are rich in vitamin C, minerals, pectin and other compounds (Guo et al., 2007). If the Se absorption ability of *C. betacea* can be improved, it will become an excellent choice for supplementing Se through food. With this in mind, the effects of IAA on the growth and Se absorption characteristics of *C. betacea* were studied to provide a reference for Se-enriched *C. betacea* production.

2 Materials and methods

2.1 Materials

The seeds of *C. betacea* were collected from a *C. betacea* tree at the Chengdu Campus of Sichuan Agricultural University (30° 71′ N, 103° 87′ E). *C. betacea* seeds were sown in substrate [perlite and vermiculite (V:V = 1:1)] and kept in an artificial climate chamber under the following conditions: 14-h days at 25 ℃, relative humidity 70%, 4,000 Lux; and 10-h nights at 20 ℃, relative humidity 90%, 0 Lux (Li et al., 2020).

2.2 Experimental design

C. betacea seedlings were cultivated in a greenhouse at the Chengdu campus of Sichuan Agricultural University from October to November 2018 under the above conditions. The strong and consistent *C. betacea* seedlings (four true leaves) were transplanted into the plastic pot (diameter × height = 15 cm × 10 cm) filled with perlite and vermiculite (V:V = 1:1) and thoroughly irrigated with deionized water. There were two *C. betacea* seedlings per pot. A volume of 100 mL Hoagland nutrient solution with a concentration of Se at 0.1 mg/L was added to the pots every 3 days to cultivate seedlings until the plants were harvested. The seedlings were sprayed with IAA seven days after transplantation. Five concentrations of IAA were used to treat the seedlings in this experiment: 0 (control), 30, 60, 90, and 120 mg/L, and there were repetitions for each treatment (two pots as a repeat). Different concentrations of IAA were sprayed evenly on the front and back surfaces of seedling leaves until the leaves began to drip. The pots were placed completely randomly. The positions of pots and the pots themselves were changed irregularly during the whole growth process to reduce the influence of marginal effects.

2.3 Determination of indicators

In November 2018, the net photosynthetic rate (Pn), transpiration rate (Tr), stomatal conductance (Gs), and intercellular CO_2 concentration (Ci) of the sixth mature leaves (counting from the top) of each plant were measured (measured within 30 min) by a Li-6400 portable photosynthesis analyzer (LI-COR Inc., Lincoln, NE, USA) from 8:00 to 10:00 in the morning. 400 μmol/mol concentration of the CO_2 was manually controlled, 25 ℃ of the temperature was maintained, and 1,200 μmol/(m^2/s) of the light intensity was controlled during photosynthesis measurement. After that, we collected the mature leaves (the fifth and sixth leaves from the top) of each plant to determine the photosynthetic pigments (chlorophyll a, chlorophyll b, total chlorophyll, and carotenoid) content, activities of antioxidant enzyme [superoxide dismutase (SOD), peroxidase (POD), and catalase (CAT)] and soluble protein content. For photosynthetic pigments content measurement, 0.200 g shredded leaves were used with the ethanol and acetone (1:1, v/v) to extract the chlorophyll a, chlorophyll b, total chlorophyll, and carotenoid content (Hao et al., 2004). To measure SOD activity, POD activity, CAT activity, and soluble protein content, fresh samples were homogenized in potassium phosphate buffer and centrifuged. The supernatant was used for determining the SOD activity, POD activity, CAT activity, and soluble protein content by the nitroblue tetrazolium photoreduction method, guaiacol colorimetric method, potassium permanganate titration methods, and Coomassie brilliant blue G-250 method, respectively (Hao et al., 2004). Later, we harvested the plants, washed, and dried to a constant weight at 75 ℃. The biomass (dry weight) of different organs was measured, respectively. Finally, the dried samples were crushed, and 0.5 g samples were weighed and digested in HNO_3/ $HClO_4$(4:1, v/v), using HCl as a reducing agent. The content of Se was determined using an iCAP 6300 inductively coupled plasma spectrometer (Thermo Scientific, Waltham, MA, USA). The shoot biomass, root/shoot ratio, Se translocation factor (TF), and Se bioconcentration factor (BCF) were calculated using the following formula: shoot biomass = stem biomass + leaf biomass; root/shoot ratio = root biomass/shoot biomass; TF = Se content in shoots/Se content in roots (Rastmanesh et al., 2010); and BCF = Se content in plants/Se concentration in culture substrate (Zhang et al., 2011).

2.4 Statistical analyses

We used a one-way analysis of variance with a least significant difference test (5% confidence level) to analyse the data using SPSS 20.0 (IBM, Chicago, IL, USA).

3 Results

3.1 Biomass

Compared to the control, the concentrations of 30 and 120 mg/L IAA had no significant differences ($P > 0.05$) on the root biomass of *C. betacea* seedlings, while the concentrations of 60 and 90 mg/L IAA increased the root biomass by 20.41% and 38.78%, respectively (Table 3-12). With the exception of the IAA concentration of 120 mg/L, sprays of 30, 60, and 90 mg/L IAA increased the stem and shoot biomass compared to those of the control, while only sprays of 30 and 60 mg/L IAA increased the biomass of leaves. Moreover, when IAA concentration was 60 mg /L, the stem, leaf, and shoot biomass of *C. betacea* seedlings reached their maximal values, which were 71.70%, 28.57%, and 38.31%, respectively, higher than the control. At 30 and 60 mg/L IAA concentrations, *C. betacea* seedlings root/shoot ratio was lower than that of the control, while at 90 and 120 mg/L, the root/shoot ratio was the opposite.

Table 3-12 Seedlings biomass of *C. betacea*

IAA concentration/ (mg/L)	Root/ (g/plant)	Stem/ (g/plant)	Leaf/ (g/plant)	Shoot/ (g/plant)	Root/shoot ratio
0	0.049±0.001c	0.053±0.002c	0.196±0.004b	0.248±0.006c	0.196±0.001b
30	0.052±0.002c	0.076±0.002b	0.245±0.008a	0.321±0.010a	0.164±0.011c
60	0.059±0.002b	0.091±0.004a	0.252±0.009a	0.343±0.013a	0.173±0.011bc
90	0.068±0.002a	0.077±0.003b	0.205±0.004b	0.282±0.007b	0.242±0.0003a
120	0.049±0.002c	0.052±0.003c	0.194±0.005b	0.247±0.007c	0.197±0.014b

Note: Different lowercase letters indicate significant differences among the treatments (least significant difference test, $P < 0.05$). Shoot biomass = stem biomass + leaf biomass; root/shoot ratio = root biomass/ shoot biomass.

3.2 Photosynthetic pigment content *C. betacea* seedlings

The 60 mg/L concentration IAA increased the chlorophyll a, chlorophyll b, and total chlorophyll content of *C. betacea* seedlings compared to the control, while other IAA concentrations had no significant differences ($P > 0.05$) on that (Table 3-13). Different IAA concentrations had no significant ($P > 0.05$) effects on the chlorophyll a/b. For carotenoid content, compared to the control, the concentrations of 30, 60, and 90 mg/L IAA increased the carotenoid content, while the concentration of 120 mg/L IAA decreased that.

Table 3-13 Photosynthetic pigment content in *C. betacea* seedlings

IAA concentration/ (mg/L)	Chlorophyll a/ (mg/g)	Chlorophyll b/ (mg/g)	Total chlorophyll/ (mg/g)	Chlorophyll a/b	Carotenoid/ (mg/g)
0	1. 35±0. 046b	0. 384±0. 013bc	1. 73±0. 059bc	3. 52±0. 001a	0. 268±0. 011c
30	1. 40±0. 052ab	0. 422±0. 022ab	1. 82±0. 074ab	3. 32±0. 054a	0. 295±0. 006b
60	1. 50±0. 055a	0. 451±0. 018a	1. 95±0. 037a	3. 34±0. 255a	0. 405±0. 015a
90	1. 39±0. 019ab	0. 423±0. 011ab	1. 82±0. 030ab	3. 30±0. 038a	0. 384±0. 00a
120	1. 30±0. 041b	0. 378±0. 016c	1. 68±0. 058c	3. 43±0. 039a	0. 232±0. 008d

Note: Different lowercase letters indicate significant differences among the treatments (least significant difference test, $P < 0.05$).

3.3 Photosynthetic characteristics of *C. betacea* seedlings

The Pn, Gs, Ci, and Tr values of *C. betacea* seedlings increased at first as the increase of IAA concentration from 0 to 120 mg/L, and they began to decrease when IAA concentration exceeded 60 mg/L (Table 3-14). Compared with the control, the IAA concentration of 30, 60, and 90 mg/L increased Pn by 71.83%, 73.37%, and 49.85%, respectively. However, the Pn did not increase significantly ($P > 0.05$) when IAA concentration was 120 mg/L. The Gs, Ci, and Tr values were the maximum at the 60 mg/L IAA concentration, followed by the concentrations of 30, 90, and 120 mg/L. Exceptionally, the Gs and Tr values of the *C. betacea* seedlings were approximately twice as high as those of the control when IAA concentration was 60 mg/L.

Table 3-14 Photosynthetic characteristics of *C. betacea* seedlings

IAA concentration/ (mg/L)	Pn/ [μmol CO_2/(m^2/s)]	Gs/ [mol H_2O/(m^2/s)]	Ci/ (μmol CO_2/mol)	Tr/ [mmol H_2O/(m^2/s)]
0	3. 23±0. 08c	0. 028±0. 001d	185. 54±8. 94c	0. 63±0. 005c
30	5. 55±0. 14a	0. 058±0. 001a	220. 38±4. 41ab	1. 23±0. 011a
60	5. 60±0. 14a	0. 060±0. 001a	221. 44±2. 49a	1. 27±0. 015a

continued table

IAA concentration/ (mg/L)	Pn/ [μmol CO_2/(m^2/s)]	Gs/ [mol H_2O/(m^2/s)]	Ci/ (μmol CO_2/mol)	Tr/ [mmol H_2O/(m^2/s)]
90	4. 84±0. 13b	0. 047±0. 002b	215. 94±3. 34ab	1. 04±0. 014b
120	3. 53±0. 12c	0. 032±0. 001c	207. 27±4. 43b	0. 65±0. 021c

Note: Different lowercase letters indicate significant differences among the treatments (least significant difference test, $P < 0.05$). Pn = net photosynthetic rate; Tr = transpiration rate; Gs = stomatal conductance; Ci = intercellular CO_2 concentration.

3.4 Antioxidant enzyme activities of *C. betacea* seedlings

C. betacea seedlings' antioxidant enzyme activity and soluble protein content all had the trend of increasing first and then decreasing with the increase in concentration of IAA (Table 3-15). Compared with the control, The SOD activity at the 30, 60, and 90 mg/L IAA concentrations were all higher, which increased by 13. 95%, 24. 18%, and 20. 68%, respectively. Compared the control, treatment with different concentrations of IAA all increased the POD activity. Among them, the POD activity treated with 60 mg/L IAA got the highest, and this difference was with the treatments of remaining concentration. With the exception of the 120 mg/L concentration, other concentrations of IAA increased the CAT activity and soluble protein content. Compared with those of the control, the concentration of 60 mg/L, IAA increased the CAT activity and soluble protein content by 54. 44% and 49.85%, respectively.

Table 3-15 Antioxidant enzyme activity of *C. betacea* seedlings

IAA concentration/ (mg/L)	SOD activity/ (U/g)	POD activity/ [U/(g/min)]	CAT activity/ [mg/(g/min)]	Soluble protein content/ (mg/g)
0	221. 89±9. 72b	1,034. 70±44. 70d	5. 97±0. 29c	92. 17±4. 51d
30	252. 84±10. 45a	1,147. 20±54. 88c	7. 04±0. 38b	108. 66±8. 89bc
60	275. 55±6. 87a	1,378. 74±33. 99a	9. 22±0. 40a	138. 12±3. 43a
90	267. 77±9. 36a	1,273. 94±37. 76b	8. 73±0. 30a	122. 17±3. 32ab
120	221. 03±7. 98b	1,266. 10±16. 30b	5. 97±0. 26c	94. 68±8. 69cd

Note: Different lowercase letters indicate significant differences among the treatments (least significant difference test, $P < 0.05$).

3.5 Se content

The Se content in various organs of *C. betacea* seedlings increased with the IAA concentration increase, reaching a maximum at 60 mg/L and then tending to decrease (Table 3-16). The concentration of 60 mg/L IAA had the highest Se content in roots, which increased by 21.14% compared with that of the control. The concentrations of 30, 90, and 120 mg/L IAA also increased the levels of Se content in the roots, but there was no significant difference ($P > 0.05$) among them. The Se content in stems was higher than that of the control only at the concentrations of 60 and 90 mg/L, which increased by 46.00% and 29.33%, respectively. With the exception of 30 mg/L IAA concentration, the Se content in leaves and shoots treated with other concentrations was higher than that the control, and the concentration of 60 mg/L IAA had the highest Se content in shoots with a 22.56% increase compared with the control. In terms of the TF, there was no significant ($P > 0.05$) difference among the treatments.

Table 3-16 Se content in *C. betacea* seedlings

IAA concentration/ (mg/L)	Root/ (mg/kg)	Stem/ (mg/kg)	Leaf/ (mg/kg)	Shoot/ (mg/kg)	TF
0	13.01±0.20c	1.50±0.015c	1.28±0.025c	1.33±0.022c	0.102±0.003a
30	13.89±0.24b	1.57±0.024c	1.34±0.047bc	1.40±0.030bc	0.101±0.0005a
60	15.76±0.21a	2.19±0.084a	1.42±0.026a	1.63±0.041a	0.103±0.004a
90	14.69±0.52b	1.94±0.064b	1.41±0.010a	1.56±0.008a	0.106±0.004a
120	13.95±0.37b	1.57±0.086c	1.40±0.031ab	1.44±0.044b	0.103±0.006a

Note: Different lowercase letters indicate significant differences among the treatments (least significant difference test, $P < 0.05$). Se content in shoot = (Se content in stem × stem biomass + Se content in leaf × leaf biomass)/shoot biomass; TF = Se content in shoot/Se content in root.

3.6 Se BCF

The Se BCF all had the trend of increase first and then decrease with the increase in concentration of IAA, with the maximum at 60 mg/L IAA concentration (Table 3-17). When the concentrations of IAA were 30, 60, 90, and 120 mg/L, the root BCFs were higher than the control. The concentration of 30 mg/L IAA had no significant effects ($P > 0.05$) on the BCFs of stem, leaf, and shoot, while the concentrations of 60, 90, and 120 mg/L IAA had higher BCFs of stem, leaf, and shoot than the control.

Table 3-17 Se BCF in *C. betacea* seedlings

IAA concentration/ (mg/L)	Root BCF	Stem BCF	Leaf BCF	Shoot BCF
0	130. 13±2. 00c	14. 97±0. 15c	12. 80±0. 25b	13. 26±0. 22c
30	138. 87±2. 36b	15. 72±0. 24c	13. 43±0. 47ab	13. 97±0. 30bc
60	157. 58±2. 11a	21. 87±0. 84a	14. 22±0. 26a	16. 26±0. 41a
90	146. 95±5. 20b	19. 39±0. 64b	14. 13±0. 10a	15. 56±0. 08a
120	139. 46±3. 69b	15. 69±0. 86c	14. 04±0. 31a	14. 39±0. 44b

Note: Different lowercase letters indicate significant differences among the treatments (least significant difference test, $P < 0.05$). BCF = Se content in plants/Se concentration in culture substrate.

4 Discussion

Some agronomic measures promote the growth and biomass accumulation of crop plants, including the plant hormone of IAA (Li et al., 2019, 2020; Wang et al., 2019; Xia et al., 2020; Yao et al., 2019; Zhang, et al., 2019). The main function of IAA is to promote the plants growth, primarily through inducing cell differentiation, division, and elongation (Zou et al., 2018). However, most studies have found that the regulation of exogenous auxin on plant growth is closely related to its concentration. Typically, a low concentration of auxin promotes growth (Wu et al., 2017), and within a certain range, the stimulatory effect increases with that of the concentration, which can decrease when it exceeds a certain concentration (Liu et al., 2018). When the concentration of auxin is too high, it inhibits the plants growth (Ye et al., 2019). In this experiment, the effects of IAA on *C. betacea* seedlings' growth are consistent with the previous studies (Liu et al., 2018; Ye et al., 2019). With the increase of the IAA concentrations, the root, stem, leaf, and shoot biomass of *C. betacea* seedlings increased first and then decreased. Different concentrations of IAA stimulates the increase in biomass compared with the control. The increase in root biomass is largely due to the formation of lateral roots and adventitious roots induced by IAA (Feng et al., 2006). Simultaneously, IAA can activate the cytoplasmic membrane of plants to pump protons out of the cells, which further acidifies and loosens the cell walls, causing the elongation of the cells and increasing the root surface area (Liu et al., 2007). When the IAA concentration was 90 mg/L, the root biomass of the seedlings was the largest, followed by 60 mg/L. IAA also promoted the other organs' biomass, which may be due to two reasons. One is that the exogenous IAA is directly absorbed and directly promotes the elongation of stem and increase in leaf area (Wu et al., 2017). The other is that the increase in roots promotes the water and mineral nutrition absorption by plants, thus contributing to the shoot growth (Liu et al., 2007). The biomass of stems, leaves and shoots all reached their maximum at an IAA concentration of 60 mg/L.

Exogenous IAA can improve the photosynthetic efficiency of plants by promoting chlorophyll synthesis (Miao et al., 2018; Hua et al., 2019; Zhao et al., 2017). In this experiment, IAA at a concentration of 60 mg/L increased *C. betacea* seedlings' photosynthetic pigment content, which could be because it promoted IAA signal conduction in the *C. betacea*, and up-regulate the expression of genes related to photosynthetic pigment synthesis, such as the genes encoding chlorophyll a and chlorophyll b binding protein, thus increasing the pigment content in leaves (Guo et al., 2016). *C. betacea* seedlings' photosynthetic pigment content as lower than the control when the concentration of IAA was 120 mg/L. This could be because IAA at this concentration inhibits the growth and reduces the overall metabolic level of plants, thus, reducing the synthesis of photosynthetic pigments to some extent. *C. betacea* seedlings' Pn, Gs, Ci, and Tr values of treated with IAA at concentrations of 30, 60, and 90 mg/L were all higher than the control. This is consistent with the result that IAA increased the concentrations of chlorophyll and carotenoid of *C. betacea* seedlings in this experiment; alternatively, some studies have found that IAA can improve the activity of RuBp carboxylase (Xiao and Li, 1987), facilitating the formation of transportable and structural substances during photosynthesis, and promoting the output of photosynthetic products (Zhang et al., 1996). These changes would improve the efficiency of light energy utilization, so this can also explain the stimulatory effect of IAA on photosynthesis of *C. betacea* seedlings.

Studies in recent years have shown that exogenous IAA also can improve the resistance and adaptability of plants, since it can improve the activities of antioxidant enzymes activities (Li et al., 2017; Liu et al., 2009; Zhang et al., 2019). In this experiment, treatments of IAA at concentrations of 30, 60, and 90 mg/L increased *C. betacea* seedlings' SOD, POD, and CAT activities. It could be because that IAA signal transduction induced the expression of genes related to antioxidant enzymes. Soluble protein is an important osmoregulatory substance in plants, which can maintain the osmotic pressure of cells and play a role in the stimulation of oxidase activity (Hua et al., 2019). The soluble protein content of seedlings treated with IAA was higher than the control, with the maximum at the concentration of 60 mg/L; this could be related to the fact that IAA can promote the biosynthesis of RNA and proteins in plants (Trewavas, 1968). Therefore, an appropriate concentration of IAA can increase *C. betacea* seedlings' antioxidant enzyme activity and soluble protein content, thus, improving its adaptability to the environment.

There are many factors affecting the Se absorption, which can be summarized in two aspects: soil and plants. Soil conditions and the Se form and content in soil are the key factors that affect the absorption of Se by plants (Ogaard et al., 2006; Zhao et al., 2007). In addition, since excessive Se amounts can cause toxic damage to plants, the tolerance of plants to Se will also affect the Se content in them (Fu et al., 2011; Shao et al., 2005). Some studies on the Se tolerance and hyperaccumulation mechanism in *Stanleya pinnata* indicate that the tolerance of Se by hyper-accumulated plants is primarily due to the increase of antioxidants and enhancement of sulfur assimilation (Freeman et al., 2010). Using the same Se concentration in this experiment, IAA significantly affected the Se content of *C. betacea* seedlings. There are some current researches on the effects of plant hormones on the characteristics of plant on Se absorption. Jasmonic acid (JA) can induce the antioxidant substances accumulation, such as glutathionem, ascorbic acid, and cysteine (Tamaoki et al., 2008), and enhance the enrichment of Se in tea leaves (Zhou et al., 2012). Melanin can effectively promote the enrichment of Se in *Plantago asiatica* L. (Liao et al., 2018). In this study, we determined that IAA concentrations of 60 and 90 mg/L increases *C. betacea* seedlings' Se content and BCF compared with the control. These results indicate that on the one hand, IAA promoted *C. betacea* seedlings' growth, and the increase in root biomass may be the reason for the increase of Se content, since the Se absorption from soil is primarily conducted by plant roots (Zhao et al., 2007; Ogaard et al., 2006). Alternatively, IAA increased the antioxidant enzyme activities and soluble protein content, thus, increasing the level of antioxidant metabolism. Therefore, IAA eventually improved the tolerance of *C. betacea* seedlings to Se, thus, promoting its absorption and accumulation. IAA had no significant effects on the TF, which indicates that IAA may have no effect on *C. betacea* seedlings' ability to transport Se from roots to shoots.

5 Conclusion

The IAA concentration of 30, 60, and 90 mg/L could promote *C. betacea* seedlings' growth by increasing its biomass, photosynthesis and tolerance. The concentration of 60 and

90 mg/L could promote the Se uptake in *C. betacea* seedlings by increasing the Se content and BCF. Therefore, the appropriate concentration of IAA could promote the growth and Se absorption of *C. betacea* seedlings, and the IAA concentration of 60 mg/L was the best choice. In future work, the focus should be on the Se absorption mechanism and Se accumulation in fruit affecting by IAA.

References

[1] BASHRI G, PRASAD S M. Exogenous IAA differentially affects growth, oxidative stress and antioxidants system in Cd stressed *Trigonella foenum-graecum* L. seedlings: toxicity alleviation by up-regulation of ascorbate-glutathione cycle [J]. Ecotoxicology and Environmental Safety, 2016, 132: 329-338.

[2] BULAK P, WALKIEWICZ A, Brzezińska M. Plant growth regulators assisted phytoextraction [J]. Biologia Plantarum, 2014, 58: 1-8.

[3] DONG X H, HE Z P, GUAN C H. Effects of IAA and zeatin introduced through secondary root on translocation and portioning of photosynthate in maize [J]. Journal of China Agricultural University, 2001, 6: 21-25.

[4] FASSLER E, EVANGELOU M W, ROBINSON B H, et al. Effects of indole-3-acetic acid (IAA) on sunflower growth and heavy metal uptake in combination with ethylene diamine disuccinic acid (EDDS) [J]. Chemosphere, 2010, 80: 901-907.

[5] FENG J, QI L W, ZHANG S Z. Review of molecular mechanism of plant rooting [J]. Biotechnology Bulletin, 2006, 22: 38-44.

[6] FREEMAN J L, TAMAOKI M, STUSHNOFF C, et al. Molecular mechanisms of selenium tolerance and hyperaccumulation in *Stanleya pinnata* [J]. Plant Physiology, 2010, 153: 1630-1652.

[7] FU D D, DUAN M L, LIANG D L, et al. Effects of selenite and selenate on growth and nutrient absorption of pakchoi [J]. Plant Nutrition and Fertilizer Science, 2011, 17: 358-365.

[8] GUO B Y, ZHOU W H, YE Q L, et al. Biological characteristics and cultivation techniques of *Cyphomandra betacea* [J]. Guangdong Agriculture Science, 2007, 34: 102-103.

[9] GUO Z X, WANG F, XIANG X. Systemic induction of photosynthesis via illumination of the shoot apex is mediated sequentially by phytochrome B, auxin and hydrogen peroxide in tomato[J]. Plant Physiology, 2016, 172:1259-1272.

[10] GUO Z, LI Z S, DAI X Y, et al. Effects of auxin on tobacco root growth and potassium uptake under low potassium stress[J]. Journal of Plant Nutrition and Fertilizers, 2019, 25:1173-1184.

[11] HAO Z B, CANG J, ZHONG X. Plant physiology experiment[M]. Harbin Institute of Technology Press, Harbin, China, 2004.

[12] HASSAN S H A, BAKAR M F A. Antioxidative and anticholinesterase activity of *Cyphomandra betacea* fruit[J]. Research Communications in Psychology, Psychiatry and Behavior, 2013, 278071.

[13] HUA Z R, LI X L. Physiological effects of exogenous IAA on seedling growth of *Scutellaria baicalensis* Georgi under salt stress[J]. Journal of Shanxi Agricultural Sciences, 2019, 47(323-328):404.

[14] KAI K, HORITA J, WAKASA K, et al. Three oxidative metabolites of indole-3-acetic acid from *Arabidopsis thaliana*[J]. Phytochemistry, 2007, 68:1651-1663.

[15] KORASICK D A, ENDERS T A, Strader L C. Auxin biosynthesis and storage forms[J]. Journal of Experimental Botany, 2013, 64:2541-2555.

[16] LI Y P, PENG Y. Improvement of oxidation resistance and osmotic regulation of white clover seedlings by exogenous auxin under polyethylene glycol stress[J]. Pratacultural Science, 2017, 34:2295-2302.

[17] LI J F, NIE Z J, ZHAO P, et al. Research progress on selenium nutrition in the soil-plant system[J]. Southern Agriculture, 2016, 47:649-656.

[18] LI H, WANG J, LIN L, et al. Effects of mutual grafting on cadmium accumulation characteristics of first post-generations of *Bidens pilosa* L. and *Galinsoga parviflora* Cav. [J]. Environmental Science and Pollution Research, 2019, 26:33228-33235.

[19] LI Z, ZHU J, WANG Y, et al. Effects of exogenous indole acetic acid on growth and cadmium accumulation of *Cyphomandra betacea* seedlings[J]. International Journal of Environmental and Analytical Chemistry, 2020, 102(3):771-779.

[20] LIAO R Y, HUANG K W, LI K Q. Effects of melatonin at different concentrations on selenium accumulation in *Plantago asiatica*[J]. Journal of Chinese Medicinal Materials, 2018, 41:1539-1542.

[21] LIU D, LIi T, YANG X, et al. Enhancement of lead uptake by hyperaccumulator plant species *Sedum alfredii* Hance using EDTA and IAA[J]. Bulletin of E0nvironmental Contamination and Toxicology, 2007, 78:280-283.

[22] LIU L, LIU Y. The indispensable microelement of human bodies-selenium[J]. Heilongjiang Science, 2014, 5(6):16-17.

[23] LIU T, SHI B S. Effects of different treatments on germination of Henbane and Gentiana seeds[J]. Journal of Hebei Normal University of Science and Technology, 2018, 32(2):44-50.

[24] LIU D, LI T, YANG X, et al. Enhancement of lead uptake by hyperaccumulator plant species *Sedum alfredii* Hance using EDTA and IAA[J]. Bulletin of Environmental Contamination and Toxicology, 2007, 78:280-283.

[25] LIU J, MA X L, SHANG X W, et al. Regulation of exogenous auxin IAA on drought and salt stress during seedling stage of spring wheat (cv. Xihan No. 2)[J]. Journal of Gansu Agricultural University, 2009, 44:47-51.

[26] LIU S C, GUO H, WANG X, et al. Effects of exogenous auxins and 6-BA on seeds germination and seedlings growth of *Amaranthus tricolo*[J]. Seed, 2018, 37:92-96.

[27] LIU Q, HUO R, LIN L, et al. Effects of different rootstocks on cadmium accumulation of grafted*Cyphomandra betacea* seedlings[J]. International Journal of Environmental and Analytical Chemistry, 2019, 99:1247-1254.

[28] MIAO Y M, JIAN X, QIAN L S, et al. Effects of IAA and GA3 on melon growth and photosynthesis under low light stress[J]. Molecular Plant Breeding, 2018, 16:2335-2340.

[29] NING Q. Effect of IAA on physiological characteristics and N, P absorption and utilization in rice seedling[D]. Xi'an University of Architecture and Technology, Shanxi, China, 2013.

[30] OGAARD A F, SOGN T A, EICH-GREATOREX S. Effect of cattle manure on selenate and selenite retention in soil[J]. Nutrient Cycling in Agroecosystems, 2006, 76:39-48.

[31] RASTMANESH F, MOORE F, KESHAVARZI B. Speciation and phytoavailability of heavy metals in contaminated soils in Sarcheshmeh area, Kerman province[J]. Bulletin of Environmental Contamination and Toxicology, 2010, 85:515-519.

[32] SHAO Z H, LIN K F, XU X Q, et al. Comparison of ecotoxic effect of selenium on

seed germination of wheat and paddy rice [J]. Chinese Journal of Ecology, 2005, 24:1440-1443.

[33] TAMAOKI M, FREEMAN J L, PILON-SMITS E A H. Cooperative ethylene and jasmonic acid signaling regulates selenite resistance in *Arabidopsis* [J]. Plant Physiology, 2008, 146:1219-1230.

[34] TREWAVAS A J. Effect of IAA on RNA and protein synthesis: effects of 3-Indolylacetic acid on the metabolism of ribonucleic acid and protein in etiolated subapical sections of *Pisum sativum* [J]. Archives of Biochemistry and Biophysics, 1968, 123:324-335.

[35] WANG L X. Research progress on physiological function and mechanism of selenium for plants [J]. Journal of Anhui Agricultural Sciences, 2010, 38(31-32):47.

[36] WANG J, WANG X, LIN L, et al. Effects of different rootstocks on the growth and cadmium-accumulation characteristics of a post-grafting generation of *Cyphomandra betacea* seedlings [J]. International Journal of Environmental and Analytical Chemistry, 2019, 101:370-378.

[37] WOODWARD A W, BARTEL B. Auxin: regulation, action, and interaction [J]. Annals of Botany, 2005, 95:707-735.

[38] WU H T, ZHAN R S, AN Z P. Effects of different concentrations of auxin on onion root tip growth [J]. Agricultural Technology and Equipment, 2017, 335:12-14.

[39] XIA H, WANG Y, LIAO M, et al. Effects of different rootstocks on cadmium accumulation characteristics of the post-grafting generations of *Galinsoga parviflora* [J]. International Journal of Phytoremediation, 2020, 22:62-68.

[40] XIAO Z F, LI M Q. Effects of indole acetic acid on photosynthesis and photorespiration of *Nicotiana tabacum* L. leaves [J]. Journal of South China Agricultural University, 1987, 8:40-48.

[41] YAO H, ZHANG F, QING M, et al. Effects of mutual grafting on the cadmium accumulation characteristics of two ecotypes of *Solanum photeinocarpum* [J]. International Journal of Phytoremediation, 2019, 21:503-508.

[42] YE Q H, WANG W, XIE Q, et al. Effect of root soaked with auxins on the growth and chlorophyll fluorescence characteristics of Chinese olive seedlings [J]. Acta Botanica Boreali-Occidentalia Sinca, 2019, 39:831-839.

[43] YUAN L J, YUAN L X, YIN X B, et al. Physiological function, deficiency and its solutions on selenium: a review [J]. Current Biotechnology, 2016, 6:396-405.

[44] ZHANG H, LUO S S, GONG C, et al. Study on the isotope kinetics of the photosynthate of flag leaf of posteloral rice and IAA's affection on it [J]. Journal of Yangzhou University (Agricultural and Life Science Edition), 1996, 17:31-36.

[45] ZHANG X F, XIA H P, LI Z A, et al. Identification of a new potential Cd-hyperaccumulator *Solanum photeinocarpum* by soil seed bank-metal concentration gradient method[J]. Journal of Hazardous Materials, 2011, 189:414-419.

[46] ZHANG Y S, LI R L, LIU Y, et al. Research progress on effects of selenium on human health[J]. Internal Medicine, 2018, 13(4):623-625, 662.

[47] ZHANG T T, ZHU Y X Y, WU Y H, et al. Effects of exogenous IAA on physiological changes and cadmium tolerance of *Trichosanthes kirilowii* Maxim. under cadmium stress[J]. Journal of Soil and Water Conservation, 2019, 33(4):335-341.

[48] ZHAO F J, LOPEZ-BELLIDO F J, GRAY C W, et al. Effects of soil compaction and irrigation on the concentrations of selenium and arsenic in wheat grains[J]. Science of the Total Environment, 2007, 372:433-439.

[49] ZHAO S H, WANG H J, WANG H B. Effect of indoleacetic acid on photosynthesis of arsenic-stressed plants with different arsenic-accumulating ability [J]. Journal of Agro-Environment Science, 2017, 36:1093-1101.

[50] ZHOU W, WEI C X, WEI W. Influence of methyl jasmonate on Se content and quality of selenium-enriched spring tea[J]. Journal of Mountain Agriculture and Biology, 2012, 31:514-518.

[51] ZOU F K, WANG Q H, ZHOU J C, et al. Auxin regulating plant growth and development: research progress[J]. Chinese Agricultural Science Bulletin, 2018, 34:34-40.

Section V Melatonin

1 Introduction

Selenium (Se) is an important component of selenoamino acids and selenoproteins, which are involved in many aspects of plant metabolism; in addition, it is an elemental nutrient required by humans to prevent diseases (Rayman, 2012). Se is found in plants mainly in the form of organic Se, and the organic Se is often used in supplement foods (Thiry et al., 2012; Chen et al., 2021). Owing to the similar chemical properties with sulfur (S), Se is accumulated by plants through the metabolism and pathways of S (Chen et al., 2021; Pilon-Smits et al., 2010); however, plants differ in how they take up S (Gupta et al., 2017). An excess of Se can cause toxicity by causing oxidative stress and the denaturation of proteins, which can cause toxicity, including stunted growth, leaf yellowing, impaired nutrient uptake, and even death (Van Hoewyk, 2013; Hasanuzzaman et al., 2020). To increase Se uptake in plants to convert to organic Se for human foods, approaches, such as grafting and the application of exogenous phytohormones, have been utilized (Huang et al., 2021; Huan et al., 2021).

Melatonin (MT) is widely present in nature, and its most widely recognized role is its participation in and regulation of the responses of plants to stress (Fan et al., 2018). As an endogenous scavenger of free radicals (Reiter et al., 2001), treatment with MT can improve the antioxidant capacity of plants and scavenge free radicals (Zou, 2019; Wang et al., 2015). Under salt stress, treatment with MT increases the content of betaine and soluble sugars to enhance tolerance (Khalid et al., 2022), and it also improves the efficiency of photosynthesis and ionic homeostasis (Zhan et al., 2019). Under heavy metal stress, treatment with MT promotes growth and increases the content of photosynthetic pigments (He et al., 2020). Treatment with MT increases the uptake of nutrients by apples under drought stress by facilitating the absorption of ions (Liang et al., 2018), and it also increases the uptake of nutrients in pistachio nuts under salt stress (Liang et al., 2018). Treatment with MT can reduce the nickel uptakes in plants under nickel stress (Altaf et al., 2021). For Se uptake, treatment with MT improves the content of Se in grapes and teas (Chen et al., 2022) and promotes the uptake of Se in the medicinal plant *Plantago asiatica L.* (Liao et al., 2018).

Thus, treatment with MT may promote Se uptake in plants. However, the action mechanism of MT on Se uptake in fruit trees has not been reported.

Cyphomandra betacea has ornamental and edible values, and its fruits are rich in vitamin C, minerals, and pectin (Suárez-Montenegro et al., 2021). However, *C. betacea* is less effective at accumulating Se (Lin et al., 2020a). Thus, increasing the content of Se in *C. betacea* can increase its commercial value. If exogenous MT is applied to *C. betacea*, its Se capacity could be improved. Therefore, the effects of exogenous MT on the Se uptake of *C. betacea* under Se stress (0.1 mg/L)(Xu et al., 2022) were investigated. And the aim of this study was to determine the best dose of MT to promote *C. betacea* growth and its Se uptake. The results of this study may provide a reference to produce *C. betacea* that is enriched in Se.

2 Materials and methods

2.1 Materials

In September 2023, the seeds were collected from a fruiting tree of *C. betacea* planted at the Chengdu Campus of Sichuan Agricultural University (30°42′ N, 103°51′ E). They were then planted in a tray filled with perlite and placed in a greenhouse with specific conditions. During the day, the greenhouse had 14 h of light at 10,000 Lux, a temperature of 25 ℃, and a relative humidity of 70%, while at night, it had 10 h of darkness at 0 Lux, a temperature of 20 ℃, and a relative humidity of 90% (Liu et al., 2021). Hoagland nutrient solution was used to irrigate each tray every 3 days until the seedlings of *C. betacea* reached approximately 10 cm in height.

The MT was obtained from Beijing Solarbio Science and Technology Co., Ltd. (Beijing, China).

2.2 Experimental design

C. betacea uniform seedlings were planted in nine-hole trays filled with perlite. Six seedlings were planted in each tray. The trays were placed in the greenhouse as described in Section 2.1. Hoagland nutrient solution containing 0.1 mg/L Se [in the form of sodium selenite (Na_2SeO_3)] was irrigated every 3 days (Xu et al., 2022). Each tray was irrigated with 100 mL. Various doses of MT (0, 50, 100, 150, and 200 μmol/L) (Xia et al., 2022) were sprayed on both sides of *C. betacea* leaves, and each tray was sprayed with 25 mL of

MT. Each treatment was repeated three times (18 for all plants in total) with one tray as a replicate. The seedlings of *C. betacea* were sprayed with the same dose of MT again after 15 days. Foil was used to limit the spray contact of the growth medium during each spraying application.

2.3 Determination of parameters

After one month from the initial treatment with MT, the third mature leaf from the top was collected to measure the levels of chlorophyll a, chlorophyll b, and carotenoid, activities of POD, SOD, and CAT, and soluble protein content. The photosynthetic pigments were analyzed using the acetone-ethanol mixed extraction method (Hao et al., 2004). The activities of POD, SOD, and CAT were assayed using the methods used in previous studies (Hao et al., 2004; Lin et al., 2020b). The content of soluble protein was determined by staining with Coomassie bright blue (Hao et al., 2004). Then, the plants were harvested and washed, and the roots were immersed in 20 mol/L EDTA-Na_2 solution for 15 min to exchange the selenite of the root surface. The samples were dried to a constant weight. The dry weights (biomass) of the roots and shoots were recorded. The samples were digested and reduced to determine the Se content using a hydride generation-atomic fluorescence spectrometry (AFS-9700; Beijing Haiguang Instrument Co., Ltd., Beijing, China) (Bao, 2000). The bioconcentration factor (BCF) of Se was calculated as the level of Se in plants/media (Hoagland nutrient solution), and the translocation factor (TF) of Se was calculated as the shoot Se content/root Se content (Rastmanesh et al., 2010).

2.4 Statistical analyses

The data of all treatments with three biological replicates was analyzed using SPSS 26.0 (IBM, Armonk, NY, USA). All the data was analyzed with a one-way analysis of variance (ANOVA) and Duncan's multiple range test. Regression analysis, Pearson's correlation analysis, principal component analysis (PCA), cluster analysis, and grey relational analysis were conducted (Lin et al., 2023a; Zhang et al., 2022b).

3 Results

3.1 Biomass

Under Se stress, with the increase in the dose of MT, the biomass increased when this dose is not higher than 150 μmol/L, and it decreased when the dose of MT higher than 150 μmol/L (Figure 3-19 and Figure 3-20). The application of MT at 50 and 100 μmol/L did not affect the root biomass, while treatments at 150 and 200 μmol/L increased it by 12% and 9%, respectively, compared with the control. All of the MT treatments increased the shoot biomass. Treatments with 50, 100, 150, and 200 μmol/L of MT increased the shoot biomass by 8%, 12%, 21%, and 15%, respectively, compared with the control.

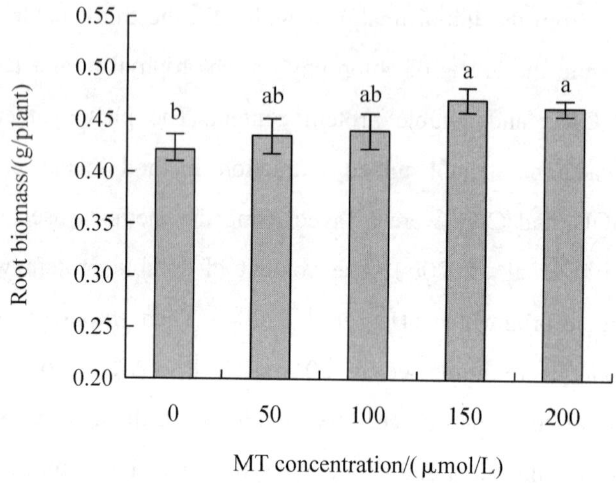

Note: Different lowercase letters indicate significant differences among the treatments (Duncan's multiple range test, $P < 0.05$).

Figure 3-9 Root biomass of *C. betacea* seedlings

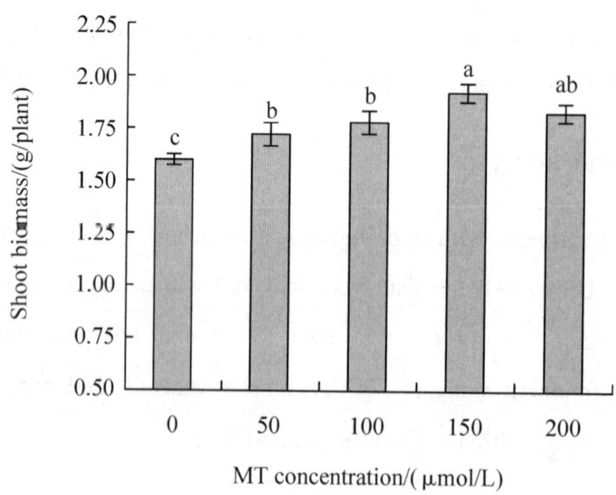

Note: Different lowercase letters indicate significant differences among the treatments (Duncan's multiple range test, $P < 0.05$).

Figure 3-20 Shoot biomass of *C. betacea* seedlings

3.2 Content of photosynthetic pigments

The content of photosynthetic pigments increased in parallel with the dose of MT under Se stress (Figure 3-21, Figure 3-22, and Figure 3-23). Treatment with MT at 50 μmol/L did not affect the content of chlorophyll a and carotenoids but increased the content of chlorophyll b. Treatments with 100, 150, and 200 μmol/L of MT increased the content of photosynthetic pigments, and these parameters reached their maximum following treatment with 150 and 200 μmol/L of MT. Treatment with 150 μmol/L of MT increased the content of photosynthetic pigments by 37%, 59%, and 32%, respectively, and increased these parameters by 36%, 57%, and 28% at 200 μmol/L, respectively, compared with the control.

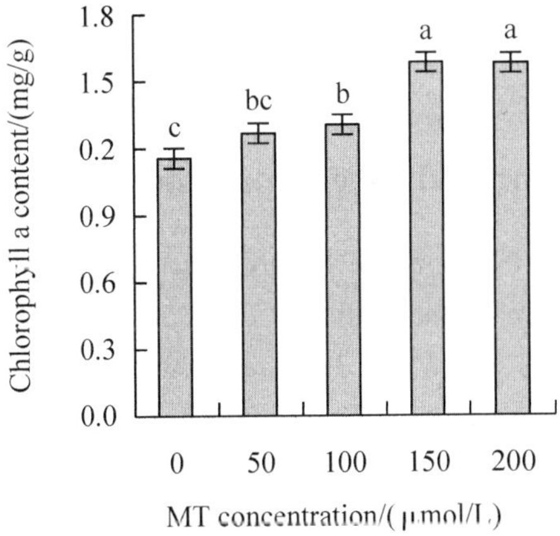

Note: Different lowercase letters indicate significant differences among the treatments (Duncan's multiple range test, $P < 0.05$).

Figure 3-21 Chlorophyll a content in *C. betacea* seedlings

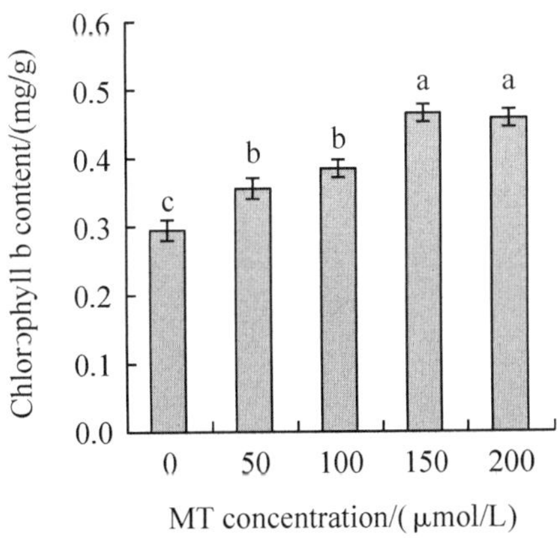

Note: Different lowercase letters indicate significant differences among the treatments (Duncan's multiple range test, $P < 0.05$). A: chlorophyll a content; B: chlorophyll b content; C: carotenoid content.

Figure 3-22 Chlorophyll b content in *C. betacea* seedlings

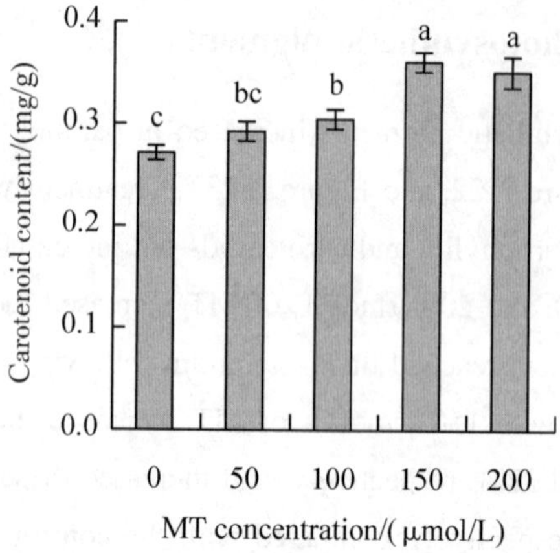

Note: Different lowercase letters indicate significant differences among the treatments (Duncan's multiple range test, $P < 0.05$).

Figure 3-23 Carotenoid content in *C. betacea* seedlings

3.3 Activities of antioxidant enzymes and content of soluble protein

The different doses of MT increased the POD activity under Se stress (Figure 3-24). Compared with the control, treatments with 50, 100, 150, and 200 μmol/L of MT increased the POD activity by 7%, 18%, 64%, and 39%, respectively. Under Se stress, treatments with 50 and 100 μmol/L of MT had no significant effects on the activities of SOD and CAT and the content of soluble protein in *C. betacea*, while treatment at 150 and 200 μmol/L increased these parameters (Figure 3-25, Figure 3-26, and Figure 3-27). Compared with the control, treatment with MT at 150 μmol/L increased the activities of SOD and CAT and the content of soluble protein by 11%, 11%, and 16%, respectively; moreover, treatment at 200 μmol/L increased these parameters by 8%, 14%, and 9%, respectively.

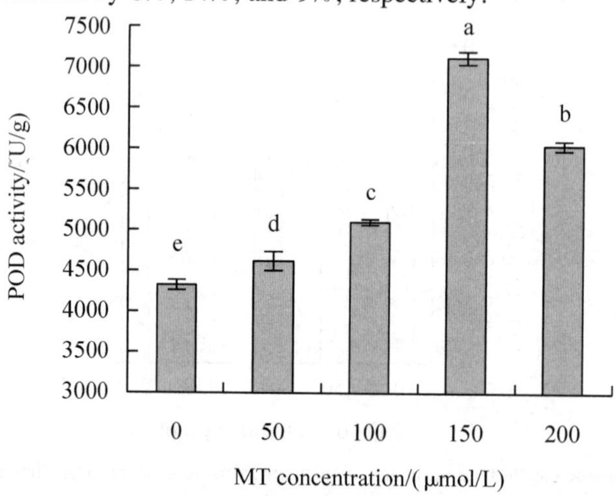

Note: Different lowercase letters indicate significant differences among the treatments (Duncan's multiple range test, $P < 0.05$).

Figure 3-24 POD activity of *C. betacea* seedlings

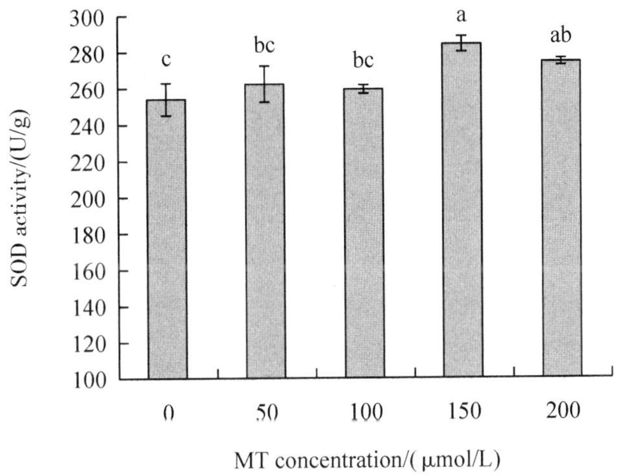

Note: Different lowercase letters indicate significant differences among the treatments (Duncan's multiple range test, $P < 0.05$).

Figure 3-25　SOD activity of *C. betacea* seedlings

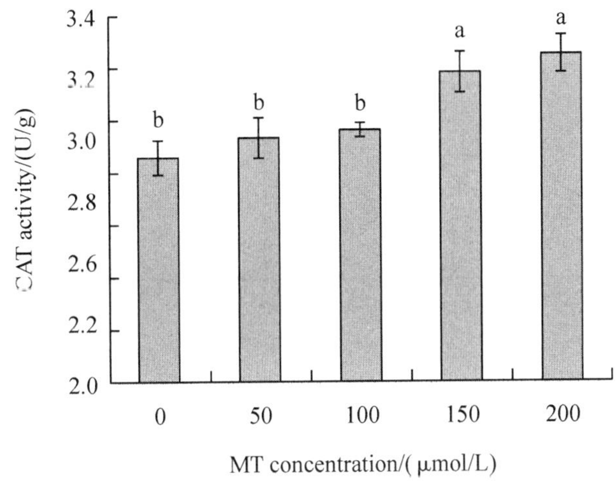

Note: Different lowercase letters indicate significant differences among the treatments (Duncan's multiple range test, $P < 0.05$).

Figure 3-26　CAT activity of *C. betacea* seedlings

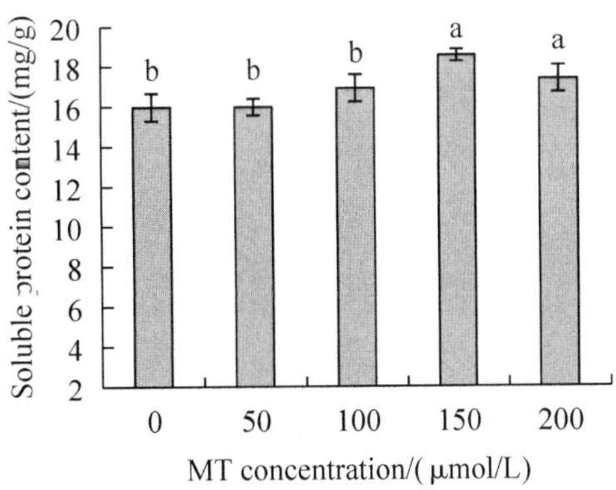

Note: Different lowercase letters indicate significant differences among the treatments (Duncan's multiple range test, $P < 0.05$).

Figure 3-27　Soluble protein content in *C. betacea* seedlings

3.4 Content and transport of Se

MT treatment increased the Se content under Se stress (Figure 3-28 and Figure 3-29). With the increase in MT dose, the Se content increased when the dose of MT is not higher than 150 μmol/L, and it decreased when the dose of MT higher than 150 μmol/L. The dose of MT had a regression relationship with the Se content and Se had the maximal content following treatment with 150 μmol/L of MT. Compared with the control, treatment with MT at 50, 100, 150, and 200 μmol/L increased the root Se content by 61%, 74%, 114%, and 47%, respectively, and increased the shoot Se content by 24%, 40%, 60%, and 31%, respectively. The different doses of MT also increased the root and shoot BCFs under Se stress (Figure 3-30 and Figure 3-31). The maximal root and shoot BCFs were obtained at 150 μmol/L of MT. However, MT treatment decreased the TF under Se stress (Figure 3-32).

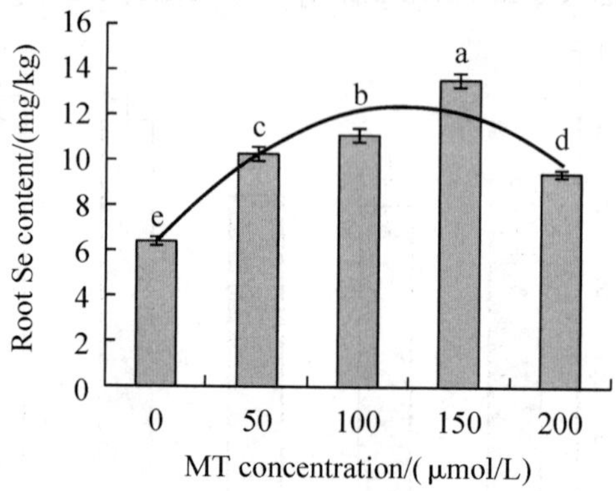

Note: Different lowercase letters indicate significant differences among the treatments (Duncan's multiple range test, $P < 0.05$).

Figure 3-28 Root Se content in *C. betacea* seedlings

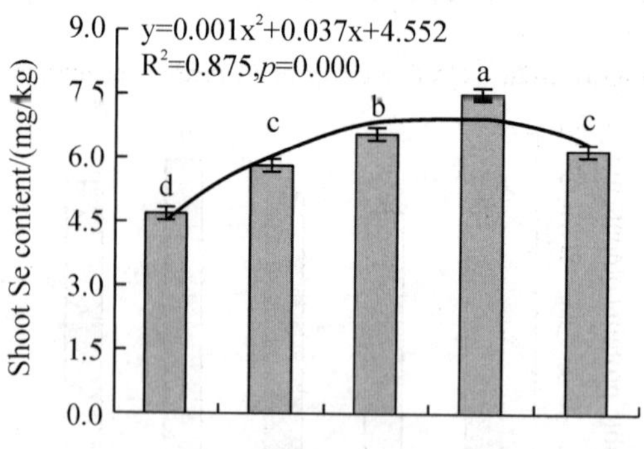

Note: Different lowercase letters indicate significant differences among the treatments (Duncan's multiple range test, $P < 0.05$).

Figure 3-29 Shoot Se content in *C. betacea* seedlings

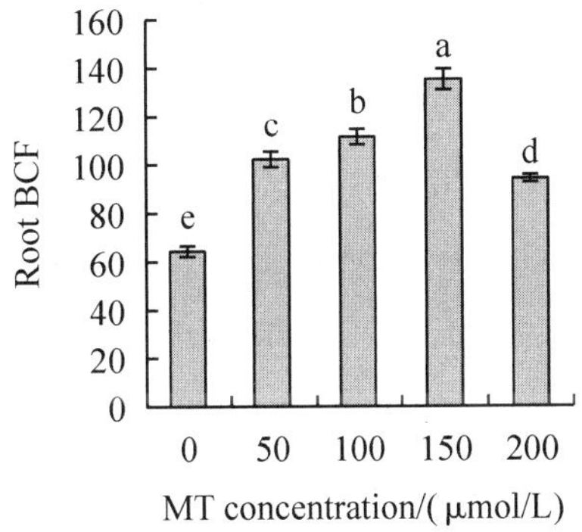

Note: Different lowercase letters indicate significant differences among the treatments (Duncan's multiple range test, $P < 0.05$).

Figure 3-30 Root bioconcentration factor (BCF) of *C. betacea* seedlings

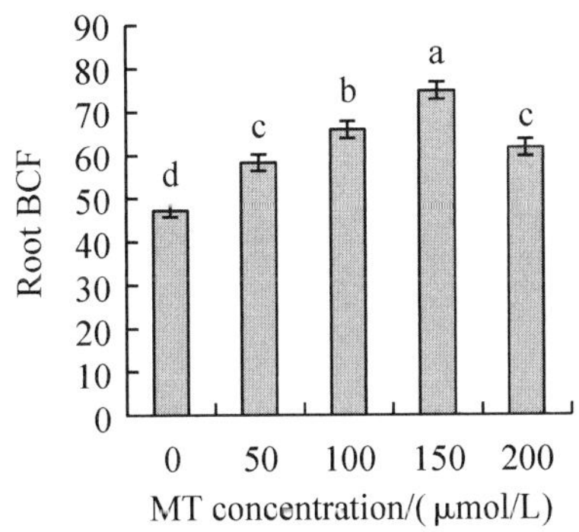

Note: Different lowercase letters indicate significant differences among the treatments (Duncan's multiple range test, $P < 0.05$).

Figure 3-31 Shoot bioconcentration factor (BCF) of *C. betacea* seedlings

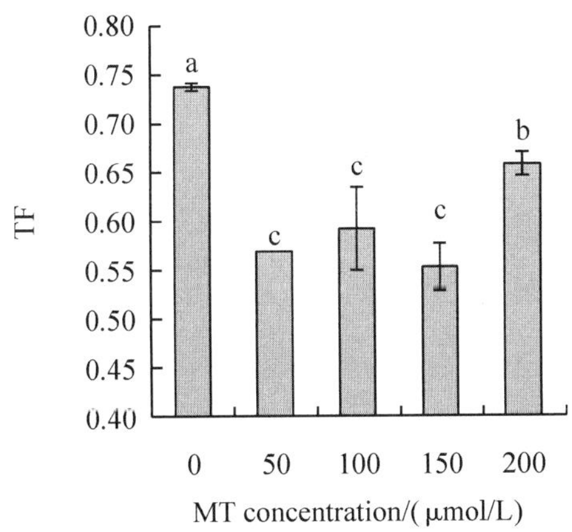

Note: Different lowercase letters indicate significant differences among the treatments (Duncan's multiple range test, $P < 0.05$).

Figure 3-32 Translocation factor (TF) of *C. betacea* seedlings

3.5 Correlation analysis

The biomass had a highly significant ($P < 0.01$)positive correlation with the content of photosynthetic pigments, antioxidant enzyme activitIES, soluble protein content, and Se content (Table 3-18). The Se content was significantly ($P < 0.01$)and positively correlated with the content of photosynthetic pigments and soluble protein and the activities of POD and SOD. The content of Se in the shoots was also significantly ($P < 0.01$) and positively correlated with the content of Se in the roots and significantly ($0.05 < P \leqslant 0.01$) and positively correlated with the activity of CAT.

Table 3-18 Correlations among the different indicators

Indicator	Root biomass	Shoot biomass	Chlorophyll a content	Chlorophyll b content	Carotenoid content	POD activity	SOD activity	CAT activity	Soluble protein content	Root se content	Shoot se content
Root biomass											
Root biomass	0.863 **										
Shoot biomass	0.935 **	0.858 **									
Chlorophyll a content	0.861 **	0.935 **	0.947 **								
Chlorophyll b content	0.935 **	0.874 **	0.994 **	0.952 **							
POD activity	0.892 **	0.907 **	0.927 **	0.905 **	0.941 **						
SOD activity	0.948 **	0.828 **	0.942 **	0.841 **	0.943 **	0.923 **					
CAT activity	0.894 **	0.777 **	0.947 **	0.898 **	0.929 **	0.860 **	0.853 **				
Soluble protein content	0.845 **	0.837 **	0.876 **	0.867 **	0.916 **	0.932 **	0.842 **	0.803 **			
Root Se content	0.683 **	0.864 **	0.628 *	0.729 **	0.682 **	0.759 **	0.699 **	0.499	0.739 **		
Shoot Se content	0.782 **	0.946 **	0.740 **	0.837 **	0.783 **	0.850 **	0.758 **	0.623 *	0.828 **	0.965 **	

Note: ** = Correlation is significant at the 0.01 level (two-tailed test). * = Correlation is significant at the 0.05 level (two-tailed test).

3.6 PCA and cluster analysis

The total variance of the two components of the PCA was 94.30%; that of PC1 was 86.37%, while that of PC2 was 7.93% (Figure 3-33). The content of Se in roots and shoots and shoot biomass were closely related. Similarly, the other parameters were also closely related. As shown by a cluster analysis (Figure 3-34), the parameters were divided into three categories when the decision distance is lower than 1.3. The first category was composed of the root biomass; activities of SOD, CAT, and POD; and the content of photosynthetic pigments. The second category was composed of the content of Se in roots and shoots and shoot biomass. The third category was composed of the soluble protein content.

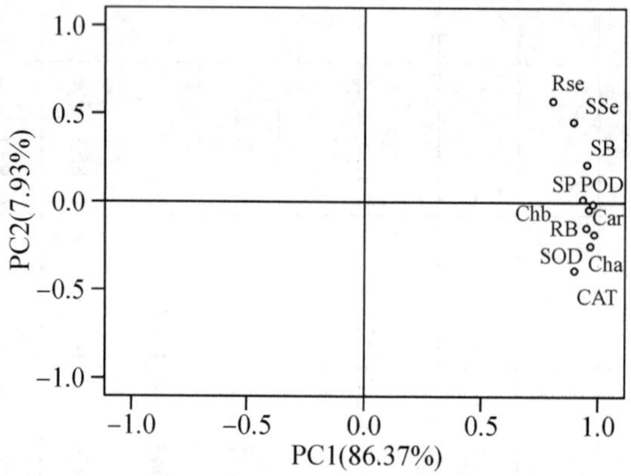

Note: RB = root biomass; SB = shoot biomass; Cha = chlorophyll a content; Chb = chlorophyll b content; Car = carotenoid content; POD = POD activity; SOD = SOD activity; CAT = CAT activity; SP = soluble protein content; RSe = root Se content; SSe = shoot Se content.

Figure 3-33 Principal component analysis (PCA)

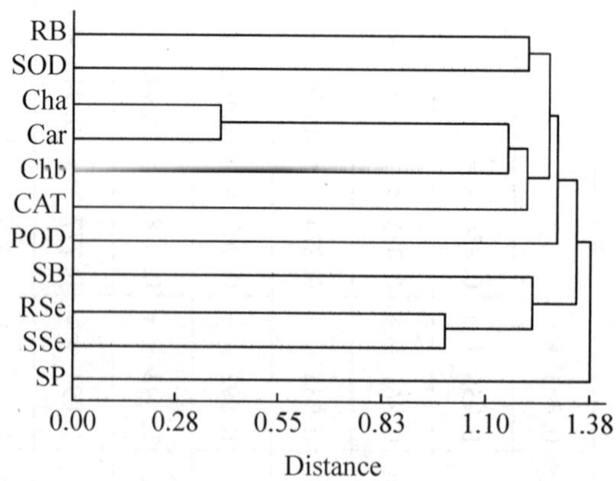

Note: RB = root biomass; SB = shoot biomass; Cha = chlorophyll a content; Chb = chlorophyll b content; Car = carotenoid content; POD = POD activity; SOD = SOD activity; CAT = CAT activity; SP = soluble protein content; RSe = root Se content; SSe = shoot Se content.

Figure 3-34 Cluster analysis

3.7 Grey and path analyses

The grey correlation coefficients of the various parameters with the shoot Se content were 0.253 ~ 0.499 (Figure 3-35). The grey correlation coefficients of the root Se content and shoot biomass were much higher than those of the other parameters. Thus, the root Se content and shoot biomass were in the closest relationships with the shoot Se content. A path analysis also showed that the root Se content and shoot biomass had the top two largest total path positive coefficients with the shoot Se content, which were much higher than those of the other parameters (Table 3-19).

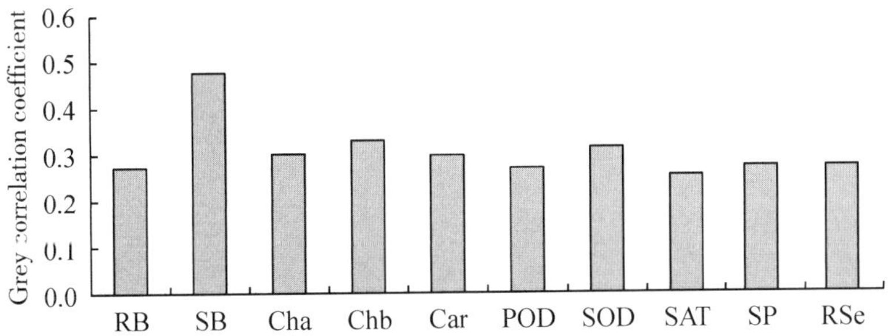

Note: RB = root biomass; SB = shoot biomass; Cha = chlorophyll a content; Chb = chlorophyll b content; Car = carotenoid content; POD = POD activity; SOD = SOD activity; CAT = CAT activity; SP = soluble protein content; RSe = root Se content.

Figure 3-35 Grey correlation coefficients of different parameters with the shoot Se content

Table 3-19 Path coefficients of different indicators with the shoot Se content

Factor	Total effect	Direct effect	Indirect effect										
			Total	X1→Y	X2→Y	X3→Y	X4→Y	X5→Y	X6→Y	X7→Y	X8→Y	X9→Y	X10→Y
X1	0. 783	0. 954	−0. 171		−0. 420	−1. 124	1. 081	0. 332	0. 690	−0. 299	−0. 542	−0. 180	0. 291
X2	0. 948	−0. 487	1. 435	0. 823		−1. 032	1. 174	0. 311	0. 701	−0. 261	−0. 471	−0. 178	0. 368
X3	0. 740	−1. 202	1. 942	0. 892	−0. 418		1. 188	0. 353	0. 717	−0. 297	−0. 574	−0. 187	0. 268
X4	0. 837	1. 255	−0. 418	0. 822	−0. 456	−1. 138		0. 338	0. 700	−0. 265	−0. 544	−0. 185	0. 310
X5	0. 781	0. 355	0. 426	0. 892	−0. 426	−1. 196	1. 194		0. 727	−0. 298	−0. 563	−0. 195	0. 291
X6	0. 850	0. 773	0. 077	0. 851	−0. 442	−1. 115	1. 136	0. 334		−0. 291	−0. 521	−0. 198	0. 323
X7	0. 758	−0. 316	1. 074	0. 904	−0. 404	−1. 133	1. 055	0. 335	0. 714		−0. 516	−0. 179	0. 298
X8	0. 622	−0. 606	1. 228	0. 853	−0. 379	−1. 139	1. 126	0. 330	0. 665	−0. 269		−0. 171	0. 212
X9	0. 829	−0. 213	1. 042	0. 807	−0. 408	−1. 054	1. 088	0. 326	0. 721	−0. 266	−0. 487		0. 315
X10	0. 966	0. 426	0. 540	0. 652	−0. 420	−0. 755	0. 914	0. 242	0. 587	−0. 221	−0. 302	−0. 157	

Note: X1 = root biomass; X2 = shoot biomass; X3 = chlorophyll a content; X4 = chlorophyll b content; X5 = carotenoid content; X6 = POD activity; X7 = SOD activity; X8 = CAT activity; X9 = soluble protein content; X10 = root Se content; Y = shoot Se content.

4 Discussion

MT can be used to protect against stress (Sun et al., 2021; Arnao et al., 2018). Treatment with MT increases the biomass of rice and sugar beet under salt stress and drought stress, respectively (Yan et al., 2021; He et al., 2023), and mitigates flooding stress in soybean (Wang et al., 2021). Treatment with MT increases the biomass of grape and *P. asiatica* under Se stress (Chen et al., 2022; Liao et al., 2018). In this study, treatment with MT increased the biomass of *C. betacea* to some extent under Se stress. It has been shown that MT can regulate the transport of indole-3-acetic acid (IAA), thus promoting plant growth (Wang et al., 2016b), and MT and IAA have synergistic or cumulative effects or both (Yang et al., 2021; Ren et al., 2019). A suitable amount of exogenous MT can act as a chemical signal to up-regulate the IAA signal transducer genes' expressions, which induces the plant to produce higher amounts of IAA and thus promotes its growth (Zhang et al., 2022a). However, high doses of MT inhibit the biosynthesis of IAA, thus inhibiting plant growth (Wang et al., 2016b). The effective dose of MT for plant growth varies from plant to plant (Arnao et al., 2017; Arnao et al., 2007).

MT can regulate the biosynthesis and degradation of chlorophyll (Yang et al., 2022). Under salt stress, treatment with MT increases the content of chlorophyll in cucumber, which suggests that treatment with MT alleviates the effect of salt stress on photosynthesis (Wang et al., 2016a); and treatment with MT mitigates the inhibition of the photosynthetic rate induced by salt treatment by the inhibiting the closure of stomata (Li et al., 2017). Under nitrogen dioxide stress, treatment with MT reduces the degradation of chlorophyll in tobacco and increases the chlorophyll content by up-regulating the chlorophyll biosynthesis-related genes' expressions (Wang et al., 2022). In this study, treatment with MT increased the photosynthetic pigment content in *C. betacea* to some degree under Se stress. This result may be related to the MT's ability to reduce the expressions of chlorophyllase, thus inhibiting the degradation of chlorophyll (Zhao et al., 2015). The other reason could be that MT can protect the integrity of the chloroplasts and thus it does not impede the biosynthesis of chlorophyll (Wu et al., 2021).

Under stress conditions, MT is important for maintaining the enzymatic antioxidant system of plants (Wu et al., 2021; Danilova et al., 2021). The mechanism of its action is that MT can act as a signaling molecule to participate in the signal transduction pathway to implement physiological functions (Arnao et al., 2018). Under salt stress, treatment with MT increases the content of soluble protein in cotton (Chen et al., 2020). Under chromium (Cr) stress, treatment with MT increases the activities of antioxidant enzymes to reduce the oxidative stress response of maize, and MT is thought to function as a multidirectional signaling molecule to promote the levels of gene expression and participate in transcription (Yang et al., 2023). Under Se stress, treatment with MT increases the activities of antioxidant enzymes in grape and *P. asiatica* and also increases their content of soluble protein (Chen et al., 2020; Liao et al., 2018). In this study, the different doses of MT increased the POD activity in *C. betacea*, and treatment with 150 and 200 μmol/L MT increased the activities of SOD and CAT and the content of soluble protein of *C. betacea* under Se stress. This effect could be due to the fact that MT acts as a strong antioxidant and can directly or indirectly scavenge reactive oxygen species (ROS) in plants under stress and thus improve the antioxidant enzyme activity (Chen et al., 2020; Ren et al., 2020). MT and its metabolites also play a role in some signaling pathways and act as signaling molecules to scavenge ROS to enhance the antioxidant capacity of plants when they are under stress (Back et al., 2021).

Treatment with MT can promote nutrient uptakes by plants (Xia et al., 2022; Xia et al., 2021). Under iron stress, treatment with MT increases the uptake of iron and redistributes it in peaches and cucumbers (Ahammed et al., 2020; Lin et al., 2023b). Under Se stress, treatment with MT promotes the Se uptakes by grape and *P. asiatica* (Chen et al., 2020; Liao et al., 2018). In this study, treatments with MT increased the Se content in *C. betacea* under Se stress. The dose of MT had a regression relationship with the Se content, indicating that MT could promote Se accumulation in *C. betacea*. Because MT can increase the content of some amino acids in plant roots (Song et al., 2021), these amino acids are secreted by plant roots and combined with inorganic Se in the soil to become organic Se in proteins through biotransformation (Dong et al., 2021). The forms of Se uptake and transport in *C. betacea* affected by MT need further study. In addition, treatment with MT also increased the BCF of *C. betacea*, while it decreased the TF of *C. betacea*. Therefore, MT could promote the Se uptake by *C. betacea* but inhibit the transport of Se from the roots to shoots. In organic Se can be directly absorbed and utilized by plants in soil, which includes selenate (SeO_4^{2-}) and selenite (SeO_3^{2-}) (Zhang et al., 2003). The inorganic Se that is absorbed by the roots is converted into selenocysteine (SeCys) or selenomethionine (SeMet), which are easily immobilized in roots, and only a small portion is transported to the aboveground parts through the xylem (Li et al., 2008). Thus, the content of Se in the roots in *C. betacea* was higher than that in shoots, which induced the decrease in TF. Moreover, the shoot Se content was positively correlated with other parameters. PCA and cluster analyses showed that the content of Se in roots and shoots and shoot biomass were classified as categories. Grey and path analyses further showed that the content of Se in roots and shoot biomass had the top two largest contributions to the content of Se in shoots. These results indicate that the content of Se in roots and shoot biomass was closely associated with the content of Se in shoots, which highlighted their significant role in promoting Se uptake by *C. betacea* under Se stress.

5 Conclusion

The growth of *C. betacea* is promoted under Se stress through treatment with MT, leading to increased biomass, photosynthetic pigment content, and antioxidant enzyme activities. Thus, MT helps to alleviate the stress caused by Se on *C. betacea*. MT treatment enhances the Se content and BCF of *C. betacea* while reducing the TF. The dose of MT used is inversely correlated with Se content, with the highest levels found at 150 μmol/L. The relationship between root Se content and shoot biomass is crucial for Se uptake. Therefore, MT treatment can support the growth of *C. betacea* and its Se absorption under Se stress. Further research should explore the application of MT in horticultural crops to enhance the production of selenium-enriched products.

References

[1] AHAMMED G J, WU M, WANG Y, et al. Melatonin alleviates iron stress by improving iron homeostasis, antioxidant defense and secondary metabolism in cucumber[J]. Scientia Horticulturae, 2020, 265:109205.

[2] AHAMMED G J, XU W, LIU A R, et al. Endogenous melatonin deficiency aggravates high temperature-induced oxidative stress in *Solanum lycopersicum* L. [J]. Environmental and Experimental Botany, 2019, 161:303-311.

[3] ALTAF M A, SHAHID R, Ren M X, et al. Melatonin mitigates nickel toxicity by improving nutrient uptake fluxes, root architecture system, photosynthesis, and antioxidant potential in tomato seedling[J]. Journal of Soil Science and Plant Nutrition, 2021, 21

(3):1842-1855.

[4] ARNAO M B, HERNANDEZ-RUIZ J. Growth activity, rooting capacity, and tropism, three auxinic precepts fulfilled by melatonin[J]. Acta Physiology Plant, 2017, 39(6):127.

[5] ARNAO M B, HERNANDEZ-RUIZ J. Melatonin and its relationship to plant hormones[J]. Annals of Botany, 2018, 121(2):195-207.

[6] ARNAO M B, HERNANDEZ-RUIZ J. Melatonin promotes adventitious and lateral root regeneration in etiolated hypocotyls of *Lupinus albus* L. [J]. Journal of Pineal Research, 2007, 42(2):147-152.

[7] BACK K. Melatonin metabolism, signaling and possible roles in plants[J]. The Plant Journal, 2021, 105(2):376-391.

[8] BAO, S D. Agrochemical analysis of soils[M]. China Agriculture Press, Beijing, China, 2000.

[9] CHEN L, LIU L, LU B, et al. Exogenous melatonin promotes seed germination and osmotic regulation under salt stress in cotton (*Gossypium hirsutum* L.)[J]. PLOS ONE, 2020, 15(1):e0228241.

[10] CHEN N, ZHAO C, ZHANG T. Selenium transformation and selenium-rich foods [J]. Food Bioscience, 2021, 40:100875.

[11] CHEN Q, LIU L, YOU X Y, et al. Effects of melatonin on growth and selenium accumulation of grape seedlings[J]. Chinese Journal of Soil Science, 2022, 53:1453-1460.

[12] DANILOVA E D, ZLOBIN I E, KUZNETSOV V V, et al. Exogenic melatonin reduces the toxic effect of polymetallic stress on barley plants[J]. Doklady Biochemistry and Biophysics, 2021, 499(1):228-232.

[13] DONG Z, XIAO Y Q, WU H. Selenium accumulation, speciation, and its effect on nutritive value of *Flammulina velutipes* (golden needle mushroom)[J]. Food Chemistry, 2021, 350:128667.

[14] EL KASSIS E, CATHALA N, ROUACHED H, et al. Characterization of a selenate-resistant *Arabidopsis* mutant. Root growth as a potential target for selenate toxicity [J]. Plant Physiology, 2007, 143(3):1231-1241.

[15] FAN J, XIE Y, ZHANG Z, et al. Melatonin: A multifunctional factor in plants[J]. International Journal of Molecular Sciences, 2018, 19(5):1528.

[16] GUPTA M, GUPTA S. An overview of selenium uptake, metabolism, and toxicity

in plants[J]. Frontiers in Plant Science, 2017, 7(1):2074.

[17]HAO Z B, CANG J, XU Z. Plant Physiology Experiment[M]. Harbin Institute of Technology Press:Harbin, China, 2004.

[18]HASANUZZAMAN M, BHUYAN M B, RAZA A, et al. Selenium in plants:Boon or bane[J]?Environmental and Experimental Botany, 2020, 178:104170.

[19] HE J, ZHUANG X, ZHOU J, et al. Exogenous melatonin alleviates cadmium uptake and toxicity in apple rootstocks[J]. Tree Physiology, 2020, 40(6):746-761.

[20]HE M M, MEI S Y, ZHAI Y N, et al. Effects of melatonin on the growth of sugar beet (*Beta vulgaris* L.) seedlings under drought stress [J]. Journal of Plant Growth Regulation, 2023, 42(8):5116-5130.

[21]HUAN Y, YANG L, LIU Q, et al. Effects of indole acetic acid on the growth and selenium absorption characteristics of *Cyphomandra betacea* seedlings[J]. Acta Physiologiae Plantarum, 2021, 43(5):74.

[22] HUANG, K; WANG, Y; WEI, X, et al. Effects of mutual grafting*Solanum photeinocarpum* from two ecosystems on physiology and selenium absorption of their offspring under selenium stress[J]. Acta Physiologiae Plantarum, 2021, 43(7):96.

[23]KAMIAB F. Exogenous melatonin mitigates the salinity damages and improves the growth of pistachio under salinity stress [J]. Journal of Plant Nutrition, 2020, 43 (10):1468-1484.

[24]KHALID M, REHMAN H M, AHMED N, et al. Using exogenous melatonin, glutathione, proline, and glycine betaine treatments to combat abiotic stresses in crops[J]. International Journal of Molecular Sciences, 2022, 23(21):12913.

[25]LI H F, MCGRATH S P, ZHAO F J. Selenium uptake, translocation and speciation in wheat supplied with selenate or selenite[J]. The Veterinary clinics of North America Food Animal Practice, 2008, 178(1):92-102.

[26]LI H, CHANG J, CHEN H, et al. Exogenous melatonin confers salt stress tolerance to watermelon by improving photosynthesis and redox homeostasis[J]. Frontiers in Plant Science, 2017, 8(3):295.

[27]LIANG B, MA C, ZHANG Z, et al. Long-term exogenous application of melatonin improves nutrient uptake fluxes in apple plants under moderate drought stress [J]. Environmental and Experimental Botany, 2018, 155:650-661.

[28]LIAO R Y, HUANG K W, LI K Q, et al. Effect of different concentrations of

melatonin on selenium accumulation in *Plantago asiatica* L. [J]. Journal of Chinese Medicinal Materials, 2018, 41:1539-1542.

[29] LIN L, LI Z, WANG J et al. 24-epibrassinolide promotes selenium uptake in grapevine under selenium stress[J]. Scientia Horticulturae, 2023a, 308:111564.

[30] LIN L, LI Z, WU C, et al. Melatonin promotes iron reactivation and reutilization in peach plants under iron deficiency[J]. International Journal of Molecular Sciences, 2023b, 24(22):16133.

[31] LIN L, SUN J, CUI T, et al. Selenium accumulation characteristics of *Cyphomandra betacea* (*Solanum betaceum*) seedlings[J]. Physiology and Molecular Biology of Plants, 2020a, 26(7):1375-1383.

[32] LIN L, WU C, JIANG W, et al. Grafting increases cadmium accumulation in the post-grafting generations of the potential cadmium-hyperaccumulator*Solanum photeinocarpum* [J]. Chemistry and Ecology, 2020b, 36(7):685-704.

[33] LIU L, HAN J, DENG L, et al. Effects of diethyl aminoethyl hexanoate on the physiology and selenium absorption of grape seedlings[J]. Acta Physiologiae Plantarum, 2021, 43(8):115.

[34] PILON-SMITS E A H, QUINN C F. Selenium metabolism in plants. In Cell Biology of Metals and Nutrients[J]. Springer:Berlin/Heidelberg, Germany, 2010, 225-241.

[35] RASTMANESH F, MOORE F, KESHAVARZI, B. Speciation and phytoavailability of heavy metals in contaminated soils in Sarcheshmeh area, Kerman Province, Iran[J]. Bulletin of Environmental Contamination and Toxicology, 2010, 85(5):515-519.

[36] RAYMAN, M P. Selenium and human health[J]. Lancet, 2012, 379:1256-1268.

[37] REITER R J, TAN D X, BURKHARDT S, et al. Melatonin in Plants[J]. Nutrition Reviews, 2001, 59, 286-290.

[38] REN J H, YE J, YIN L, et al. Exogenous melatonin improves salt tolerance by mitigating osmotic, ion, and oxidative stresses in maize seedlings[J]. Agronomy, 2020, 10(5):663.

[39] REN S X, RUTTO L, KATUURAMU D. Melatonin acts synergistically with auxin to promote lateral root development through fine tuning auxin transport in *Arabidopsis* thaliana[J]. PLOS ONE, 2019, 14(8):e0221687.

[40] SONG Y, WANG B, QIU D, et al. Melatonin enhances metallic oxide nanoparticle stress tolerance in rice via inducing tetrapyrrole biosynthesis and amino acid metabolism[J]. Environmental Science:Nano, 2021, 8(8):2310-2323.

[41] SUáREZ-MONTENEGRO Z J, BALLESTEROS-VIVAS D, GALLEGO R, et al. Neuroprotective potential of tamarillo (*Cyphomandra betacea*) epicarp extracts obtained by sustainable extraction process[J]. Frontiers in Nutrition, 2021, 8:769617.

[42] SUN C L, LIU L J, WANG L X, et al. Melatonin: A master regulator of plant development and stress responses [J]. Journal of Integrative Plant Biology, 2021, 63 (1):126-145.

[43] THIRY C, RUTTENS A D, TEMMERMAN L, et al. Current knowledge in species-related bioavailability of selenium in food [J]. Food Chemistry, 2012, 130 (4):767-784.

[44] VAN HOEWYK D. A tale of two toxicities: Malformed selenoproteins and oxidative stress both contribute to selenium stress in plants[J]. Annals of Botany, 2013, 112 (6):965-972.

[45] WANG L Y, LIU J L, WANG W X, et al. Exogenous melatonin improves growth and photosynthetic capacity of cucumber under salinity induced stress[J]. Photosynthetica, 2016a, 54(1):19-27.

[46] WANG P, SUN X, WANG N, et al. Melatonin enhances the occurrence of autophagy induced by oxidative stress in *Arabidopsis* seedlings [J]. Journal of Pineal Research, 2015, 58(4):479-489.

[47] WANG Q, AN B, WEI Y, et al. Melatonin regulates root meristem by repressing auxin synthesis and polar auxin transport in *Arabidopsis* [J]. Frontiers in Plant Science, 2016b, 7(12):1882.

[48] WANG X, LI F, CHEN Z Y, et al. Proteomic analysis reveals the effects of melatonin on soybean root tips under flooding stress [J]. Journal of Proteomics, 2021, 232:104064.

[49] WANG Y, GUO D, WANG J, et al. Exogenous melatonin alleviates NO_2 damage in tobacco leaves by promoting antioxidant defense, modulating redox homeostasis, and signal transduction[J]. Journal of Hazardous Materials, 2022, 424:127265.

[50] WU C, CAO S, XIE K, et al. Melatonin delays yellowing of broccoli during storage by regulating chlorophyll catabolism and maintaining chloroplast ultrastructure [J]. Postharvest Biology and Technology, 2021, 172:111378.

[51] XIA H, SHEN Y, DENG H, et al. Melatonin application improves berry coloration, sucrose synthesis, and nutrient absorption in 'Summer Black' grape[J]. Food Chemistry, 2021, 356:129713.

[52] XIA H, YANG C, LIANG Y, et al. Melatonin and arbuscular mycorrhizal fungi synergistically improve drought toleration in kiwifruit seedlings by increasing mycorrhizal colonization and nutrient uptake[J]. Frontiers in Plant Science, 2022, 13:1073917.

[53] XU Y, ZHANG L, WANG J, et al. Gibberellic acid promotes selenium accumulation in *Cyphomandra betacea* under selenium stress[J]. Frontiers in Plant Science, 2022, 13, 968768.

[54] YAN F, WEI H, DING Y, et al. Melatonin regulates antioxidant strategy in response to continuous salt stress in rice seedlings[J]. Plant Physiology and Biochemistry, 2021, 165:239-250.

[55] YANG L, YOU J, LI J Z, et al. Melatonin promotes*Arabidopsis* primary root growth in an IAA-dependent manner [J]. Journal of Experimental Botany, 2021, 72 (15):5599-5611.

[56] YANG S, ZHAO Y, QIN X, et al. New insights into the role of melatonin in photosynthesis[J]. Journal of Experimental Botany, 2022, 73(17):5918-5927.

[57] YANG X X, REN J H, LIN X Y, et al. Melatonin alleviates chromium toxicity in maize by modulation of cell wall polysaccharides biosynthesis, glutathione metabolism, and antioxidant capacity[J]. International Journal of Molecular Sciences, 2023, 24(3816):3816.

[58] ZHAN H, NIE X, ZHANG T, et al. Melatonin: A small molecule but important for salt stress tolerance in plants [J]. International Journal of Molecular Sciences, 2019, 20 (3):709.

[59] ZHANG M, GAO C, XU L, et al. Melatonin and indole-3-acetic acid synergistically regulate plant growth and stress resistance[J]. Cells, 2022a, 11(20):3250.

[60] ZHANG R, LIU Q, XU X, et al. An amino acid fertilizer improves the emergent accumulator plant*Nasturtium officinale* R. Br. phytoremediation capability for cadmium-contaminated paddy soils[J]. Frontiers in Plant Science, 2022b, 13:1003743.

[61] ZHANG Y L, PAN G P, CHEN J, et al. Uptake and transport of selenite and selenate by soybean seedlings of two genotypes[J]. Plant and Soil, 2003, 253(2):437-443.

[62] ZHAO H, SU T, HUO L, et al. Unveiling the mechanism of melatonin impacts on maize seedling growth: Sugar metabolism as a case[J]. Journal of Pineal Research, 2015, 59 (2):255-266.

[63] ZOU, J N. Effects of exogenous melatonin on photosynthesis and growth of soybean under drought stress [D]. Heilongjiang Bayi Agricultural University, Daqing, China, 2019.

Section Ⅵ Abscisic acid

1 Introduction

Selenium (Se)has biological functions such as antioxidant and immune enhancement for human body (Jia et al., 2021). Humans obtain less than 40 μg of Se a day from the diet, and Se poisoning occurs after intake of concentration above 400 μg a day (El-Ramady et al., 2015; Supriatin et al. 2015). Fruits and vegetables are the most abundant sources of Se through diet compared with other more expensive food products like meat and seafood (Hossain et al., 2021). Se biofortification of fruits and vegetables is an effective way of supplementing Se through diet. However, because of low Se concentration in most areas of soils (0.1 ~ 0.2 mg/kg), the Se content in common fruits and vegetables are lower than 10 μg/kg, which is lower than in meat and seafood (more than 100 μg/kg)(Shi, 1995; Liu et al., 2005). It can be achieved through agronomic and genetic breeding methods that increase the accumulation of Se in crops. Crops can absorb and convert soil inorganic Se into the organic Se needed by the human body (Jiang et al., 2016; Puccinelli et al., 2017). Common Se biofortification methods include applying the inorganic Se fertilizers, such as selenate and selenite (Radawiec et al., 2021; Zafeiriou et al., 2022), or organic Se fertilizers, such as Se-enriched straw, Se yeast, and Se amino acids (Kikkert et al., 2013; Kaur et al., 2018; Zhou et al., 2021). Se fertilization is a very common Se biofortification strategy; however, this approach is expensive with low short-term efficiency. It also threatens environmental pollution (Zhang, 2018; Pu, 2021). Therefore, it is necessary to explore new technologies or measures for improving Se absorption of crops.

Various approaches have been applied to improve the Se absorption by crops, including intercropping, grafting, and application of phytohormones (Luo, 2019; Huan et al., 2021; Pan et al., 2021). Exogenous phytohormones have been extensively used in crop production to promote crop growth and development (Peleg et al., 2011). Abscisic acid (ABA) is a hemiterpenoid phytohormone that plays an essential role in regulating seed germination, seed dormancy, and plant growth and development (Cao et al., 2017; Zhu et al., 2017; Xu et al., 2020; Gupta et al., 2022). As a stress hormone, ABA plays critical roles in plant growth under stress conditions (Cutler et al., 2010; Raghavendra et al., 2010). Under stress condition, ABA accumulates rapidly in plants and transports to the stress site (Boursiac et al., 2013). ABA is predominantly produced in vascular tissues of plants and its signaling relies on the transport of ABA in synthesis cells to the target cells (Boursiac et al., 2013; Kuromori et al., 2018). For example, exogenous ABA increases the content of relative water and chlorophyll in tomato, and promotes the vegetative growth of tomato under water deficit (Ramasamy et al., 2022). Under cadmium (Cd) stress, exogenous ABA increases the antioxidant enzyme activity in stems of *Brassica campestris L.* ssp. *chinensis* var. *perperea Hort.*, directing more Cd accumulation in stems to alleviate Cd stress (Shen et al., 2017). In Cd-contaminated soil, exogenous ABA increases the biomass (dry weight) of potential Cd-hyperaccumulator *Solanum photeinocarpum* and decreased the biomass (dry weight) of Cd-hyperaccumulator *Bidens pilosa* (Liu et al., 2017; Wang et al., 2016). Regarding the nutrient uptake by crops, exogenous ABA increases the content of nitrogen, potassium, magnesium, calcium, and trace elements (iron, zinc, manganese) in grapes under salt stress (Karimi et al., 2021). Exogenous ABA alleviates iron deficiency symptoms in apple seedlings by promoting the root development and expression of iron transport-related genes to maximize the iron distribution in roots and shoots (Zhang et al., 2020). Collectively, the above reports show that ABA promotes plant growth under abiotic stress by improving plant stress resistance and nutrient uptake. Therefore, we hypothesized that exogenous ABA could improve the Se uptake of fruit tree under Se stress.

Cyphomandra betacea (*Solanum betaceum*) is a fruit tree, and its fruits are rich in minerals, vitamin C, and other nutrients, in the human diet (Zhang et al., 2005; Dong et al., 2022). However, the previous studies showed that Se accumulation capacity of *C. betacea* is low, and the concentration of 0.1 mg/L Se (Na_2SeO_3) in aqueous solution may cause stress to *C. betacea* seedlings (Liu et al., 2019; Lin et al., 2020). Thus, improving the Se accumulation capacity of *C. betacea* could improve the Se levels in human diet to some degree. In this study, the effects of different ABA concentrations on the Se uptake by *C. betacea* seedlings were studied under Se stress. The aim of this study was to determine the optimum ABA concentration for Se accumulation in *C. betacea* under Se stress. This finding could provide a reference for producing Se-enriched *C. betacea*.

2 Materials and methods

2.1 Materials

In September 2021, the mature seeds were collected fromthe mature fruits of *C. betacea* tree planting at the Chengdu Campus of Sichuan Agricultural University (30°42′N, 103°51′E), and air-dried. The seeds were sown in a 50-hole plug tray (54 cm length × 28 cm width × 4 cm height)containing perlite (with 4 ~ 8 mm diameter, sterilized by an autoclave), and placed in a greenhouse with the following conditions: 14-h at 25 ℃, 70% relative humidity, and 10,000 Lux for daytime; and 10-h at 20℃, 90% relative humidity, and 0 Lux for nighttime (Liu, 2021). Each plug tray was irrigated with 1,000 mL distilled water every three days before seeds germination. After seeds germination, each plug tray was irrigated with 1000 mL half strength nutrient Hoagland solution every three days.

2.2 Experimental design

When *C. betacea* seedlings were 5 cm in height (one-month post-emergence), four seedlings with uniform growth were transplanted into each nine-hole plug tray (14.5 cm length × 14.5 cm width × 10 cm height) filled with perlite (with 4 ~ 8 mm diameter, sterilized by an autoclave). The plug trays were placed in the same greenhouse under the above conditions. Each plug tray was irrigated with 100 mL Hoagland nutrient solution containing 0. 1 mg/L Se (Na_2SeO_3)every three days (Liu et al., 2019; Li et al., 2022b). In addition, ABA solutions (0, 1, 5, 10 and 20 μmol/L)(Wang et al., 2016; Liu et al., 2017) were sprayed on both sides of the plant leaves (25 mL per plug tray)and sprayed again after 15 days. The surface of the leaves was condensed into water droplets without shedding any drops of the solution. Each treatment had six replicates (one plug tray as one replicate). Three replicates were used to determinate the physiological parameters, and the other three replicates were used to determinate the biomass and Se content.

2.3 Analysis of parameters

After one month of seedling growth (from the first ABA treatment) in Se treatment condition, the mature leaves of each plant with three replicates in each treatment were collected to do further analysis. To determine the photosynthetic pigments (chlorophyll a, chlorophyll b, and carotenoid) content, the fresh leaf samples were soaked in ethanol and acetone (1:1, v/v) mixture solution (Liu et al., 2022). Then, the extraction solution was determined at 663, 645, and 470 nm wavelengths using a spectrophotometer (Summit, Shanghai, China) as described by Hao et al. (2004).

Also, the mature fresh leaf samples were grounded and homogenized in 0.05 M potassium phosphate extraction buffer (pH 7.0, containing 1 mM EDTA) to extract the enzyme at 4 ℃ (Liu et al., 2022). After the homogenate was centrifuged (11,000 × g for 20 min), the supernatant was used for determining the activities of antioxidant enzymes including the superoxide dismutase (SOD) by the nitroblue tetrazole method, peroxidase (POD) by the guaiacol colorimetric method, and catalase (CAT) by the potassium permanganate titration method, and determining the soluble protein content by Bradford method, proline content by the sulfosalicylic acid method, and malondialdehyde (MDA) content by the thiobarbituric acid method, respectively, as described by Hao et al. (2004) and Liu et al. (2022). The relative conductivity was determined by the electric conductometer as described by Hao et al. (2004).

Subsequently, the whole plants of other three replicates in each treatment were harvested. The root and shoot of each plant were separated, washed, and dried in an oven at 80 ℃ to obtain a constant weight (Li et al., 2022b). The root and shoot biomass (dry weight) were measured using an electronic balance, and the root/shoot ratio (root biomass/shoot biomass) was calculated.

The dry plant samples were finely ground and digested in the nitrate acid and perchloric acid (4:1, v/v) solution at 200 ℃, and then reduced by the hydrochloric acid to determine the total Se content using hydride generation atomic fluorescence spectrometry (AFS-9700, Beijing Haiguang Instrument Co., Ltd., Beijing, China) (Li et al., 2022b). The translocation factor (TF: Se content in shoots/roots) was calculated as described previously (Rastmanesh et al., 2010).

2.4 Statistical analyses

Statistical analyses were performed using SPSS 20.0 (IBM, Chicago, IL, USA). Data with three replicates as subjected to normal distribution and homogeneity test before one-way analysis of variance, followed by the Duncan's multiple range test ($P < 0.05$). A regression analysis was used to analyze the relationship between ABA concentration and biomass or Se content. Pearson's correlation was used to analyze the correlation relationship. Grey relational analysis was conducted according to the reports of Wang (2019) and Zhang et al. (2022).

3 Results

3.1 Biomass and root/shoot ratio of *C. betacea* seedlings

The lower concentration of ABA (< 20 μmol/L) decreased the root and shoot biomass of *C. betacea* seedlings, while the higher concentration (20 μmol/L) increased the biomass (Figure 3-36 and Figure 3-37). Compared with the control, the ABA concentration at 20 μmol/L increased the root biomass by 20.47%, while the 1, 5, and 10 μmol/L ABA either decreased or had no significant effect on the biomass. Meanwhile, 20 μmol/L ABA increased the shoot biomass by 5.34% compared with the control, while lower concentrations of ABA decreased the biomass relatively to the control. Regression analysis showed that the root biomass had a parabolic relationship with ABA concentration (x: ABA concentration; y: root biomass; $y = 0.002x^2 - 0.020x + 0.609$, $R^2 = 0.775$, $P = 0.000$). Also, the shoot biomass had a parabolic relationship with ABA concentration (x: ABA concentration; y: shoot biomass; $y = 0.019x^2 - 0.312x + 6.133$, $R^2 = 0.713$, $P = 0.001$). Moreover, ABA at 1, 10, and 20 μmol/L increased the root/shoot ratio of *C. betacea* seedlings, while 5 μmol/L ABA had no significant effect (Figure 3-38).

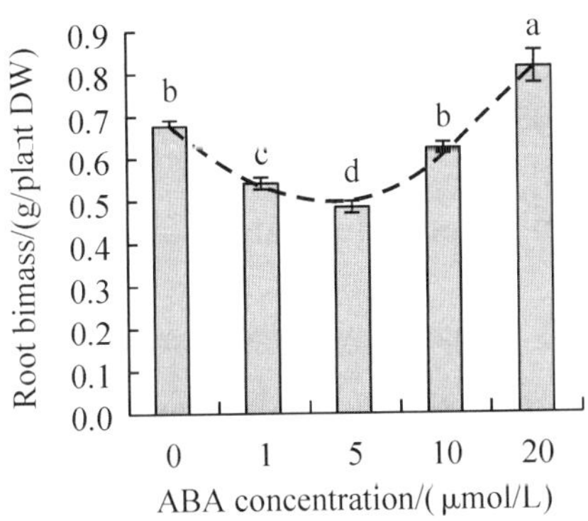

Note: Different lowercase letters indicate significant differences among the treatments (Duncan's multiple range test, $P < 0.05$).

Figure 3-36 Root biomass of *C. betacea* seedlings

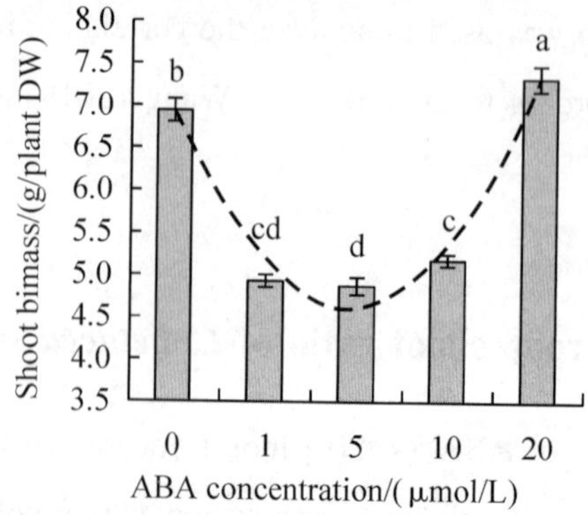

Note: Different lowercase letters indicate significant differences among the treatments (Duncan's multiple range test, $P < 0.05$).

Figure 3-37 Shoot biomass of *C. betacea* seedlings

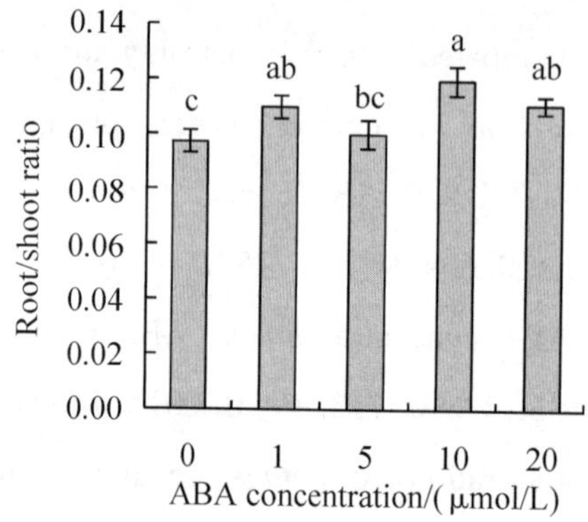

Note: Different lowercase letters indicate significant differences among the treatments (Duncan's multiple range test, $P < 0.05$).

Figure 3-38 Root/shoot ratio of *C. betacea* seedlings

3.2 Photosynthetic pigment content in *C. betacea* seedling leaves

The content of chlorophyll a, chlorophyll b, and carotenoid in *C. betacea* seedling leaves had the same response as the biomass following exogenous ABA treatment (Table 3-20). Compared with the control, 20 μmol/L ABA increased the content of chlorophyll a, chlorophyll b, and carotenoid by 17.30%, 31.76%, and 7.87%, respectively, while the lower concentrations (1, 5, and 10 μmol/L) either decreased or exerted no significant effects. Notably, the chlorophyll a/b ratio of *C. betacea* seedling leaves treated with different ABA concentrations were lower than that of the control (Table 3-20).

Table 3-20 Photosynthetic pigment content in *C. betacea* seedlings

ABA concentration/ (μmol/L)	Chlorophyll a/ (mg/g)	Chlorophyll b/ (mg/g)	Chlorophyll a/b	Carotenoid/ (mg/g)
0	2.046±0.062b	0.731±0.014b	2.799±0.032a	0.408±0.006b
1	1.752±0.020c	0.695±0.015bc	2.520±0.083b	0.371±0.018c
5	1.396±0.023d	0.638±0.021c	2.189±0.035c	0.353±0.009c
10	1.762±0.042c	0.696±0.030bc	2.532±0.050b	0.378±0.011bc
20	2.400±0.012a	0.963±0.033a	2.493±0.097b	0.440±0.014a

Note: Different lowercase letters indicate significant differences among the treatments (Duncan's multiple range test, $P < 0.05$).

3.3 Antioxidant enzyme activity of *C. betacea* seedling leaves

Compared with the control, the 5, 10, and 20 μmol/L ABA decreased the SOD activity of *C. betacea* seedling leaves by 11.43%, 26.01%, and 26.65%, respectively, and increased the POD activity by 35.70%, 83.32%, and 83.78%, respectively, while 1 μmol/L ABA had no significant effect on these two parameters (Table 3-21). ABA at 1, 5, and 10 μmol/L had no significant effect on the CAT activity in leaves, while 20 μmol/L ABA decreased the CAT activity by 4.31%, relatively to the control.

Table 3-21 Antioxidant enzyme activity of *C. betacea* seedlings

ABA concentration/ (μmol/L)	SOD activity/ (U/g)	POD activity/ [U/(g/min)]	CAT activity/ [mg/(g/min)]
0	303.0±6.39a	7,488±68c	5.286±0.107a
1	306.1±9.55a	7,589±36c	5.289±0.081a
5	268.4±6.61b	10,161±172b	5.239±0.061ab
10	224.2±5.06c	13,727±331a	5.185±0.043ab
20	222.2±7.01c	13,762±252a	5.058±0.100b

Note: Different lowercase letters indicate significant differences among the treatments (Duncan's multiple range test, $P < 0.05$).

3.4 Osmotic regulator content, MDA content, and relative conductivity of *C. betacea* seedling leaves

Compared with the control, the 1, 5, 10, and 20 μmol/L concentrations of ABA decreased the soluble protein content in *C. betacea* seedling leaves by 17.14%, 39.39%, 62.

46%, and 63. 16%, respectively, while the 5, 10, and 20 μmol/L concentrations of ABA increased the proline content by 36. 98%, 66. 55%, and 186. 49%, respectively (Table 3-22). At 5, 10, and 20 μmol/L, ABA decreased the MDA content by 7.17%, 9.41%, and 12.54%, respectively, compared with the control, while 1 μmol/L ABA had no significant effect. Furthermore, ABA at 10 and 20 μmol/L decreased the relative conductivity by 13.19% and 17. 45%, respectively, compared with the control, while 1 and 5 μmol/L ABA had no significant effect.

Table 3-22 Osmotic regulator content, MDA content, and relative conductivity of *C. betacea* seedling leaves

ABA concentration/ (μmol/L)	Soluble protein content/ (mg/g)	Proline content/ (μg/g)	MDA content/ (μmol/kg)	Relative conductivity/ (μs/cm)
0	18. 34±0. 50a	275. 3±7. 50d	2. 870±0. 06a	37. 14±0. 86a
1	15. 20±0. 48b	277. 4±11. 17d	2. 819±0. 10ab	36. 10±0. 76a
5	11. 12±0. 38c	377. 1±12. 87c	2. 664±0. 08bc	35. 99±0. 14a
10	6. 88±0. 22d	458. 5±19. 09b	2. 600±0. 05c	32. 24±1. 00b
20	6. 76±0. 21d	788. 7±29. 27a	2. 510±0. 04c	30. 66±0. 92b

Note: Different lowercase letters indicate significant differences among the treatments (Duncan's multiple range test, $P < 0.05$).

3.5 Se content and translocation factor (TF) of *C. betacea* seedlings

The root and shoot Se content in *C. betacea* seedlings increased with the increase of ABA concentration (Figure 3-39 and Figure 3-40). Compared with the control, 10 and 20 μmol/L ABA increased the root Se content by 12. 31% and 76. 64%, respectively, while 1, 5, 10, and 20 μmol/L ABA increased the shoot Se content by 27. 19%, 34. 57%, 46. 96%, and 55. 83%, respectively. Regression analysis showed that the root Se content had a parabolic relationship with ABA concentration (x: ABA concentration; y: root Se content; $y = 0.013x^2 - 0.067x + 5.556$, $R^2 = 0.990$, $P = 0.000$). On the other hand, the shoot Se content had a linear function relationship with ABA concentration (x: ABA concentration; y: shoot Se content; $y = 0.084x + 4.353$, $R^2 = 0.731$, $P = 0.000$). Also, ABA at 1, 5, and 10 μmol/L increased the TF of *C. betacea* seedlings, while 20 μmol/L ABA decreased the TF, compared with the control (Figure 3-41).

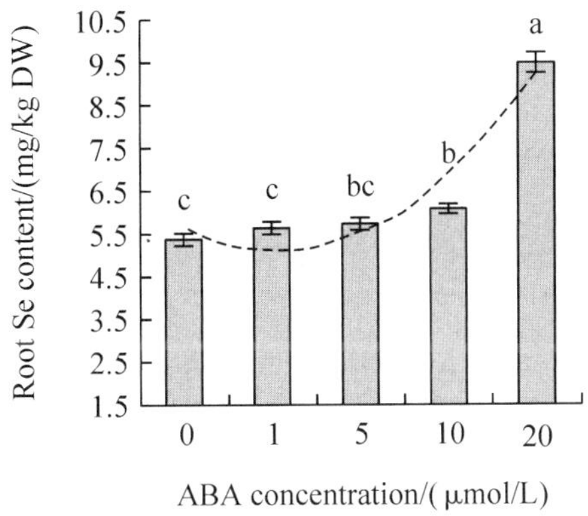

Note: Different lowercase letters indicate significant differences among the treatments (Duncan's multiple range test, $P < 0.05$).

Figure 3-39 Root Se content in *C. betacea* seedlings

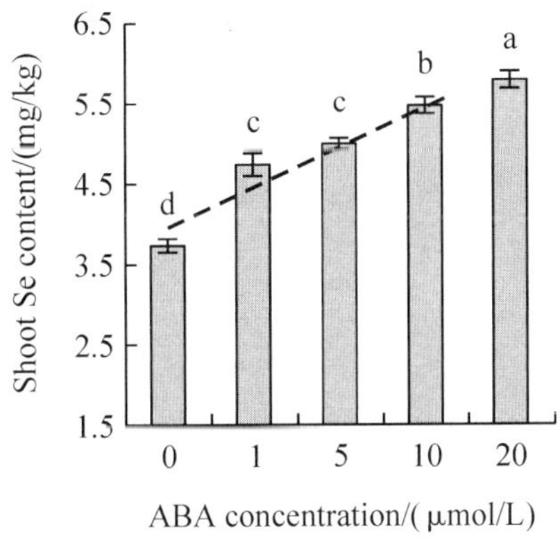

Note: Different lowercase letters indicate significant differences among the treatments (Duncan's multiple range test, $P < 0.05$).

Figure 3-40 Shoot Se content in *C. betacea* seedlings

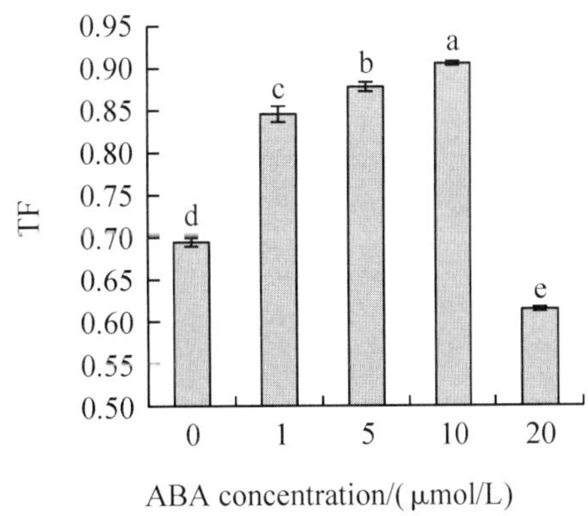

Note: Different lowercase letters indicate significant differences among the treatments (Duncan's multiple range test, $P < 0.05$).

Figure 3-41 Translocation factor (TF) of *C. betacea* seedlings

3.6 Correlation and grey relational analyses

To further analyze the relationship of different parameters with the shoot Se content, correlation and grey relational analyses were used in this study. Correlation analysis showed that the shoot total Se content showed a positive correlation with the root/shoot ratio, POD activity, proline content, and root Se content, and showed a negative correlation with the chlorophyll a/b, SOD activity, CAT activity, soluble protein content, MDA content, and relative conductivity (Table 3-23). Grey relational analysis showed that the biomass, root Se content, photosynthetic pigment content, antioxidant enzyme activity, osmotic regulation substance content, MDA content, and relative conductivity were all correlated with the shoot Se content (Figure 3-42). The top two grey correlation coefficients were POD activity and proline content. So, the POD activity and proline content exhibited the closest relationship with the shoot Se content.

Table 3-23　Correlations among the different indicators

Indicator	Root biomass	Shoot biomass	Root/shoot ratio	Chlorophyll a content	Chlorophyll b content	Chlorophyll a/b	Carotenoid content	SOD activity	POD activity	CAT activity	Soluble protein content	Proline content	MDA content	Relative conductivity	TF	Root Se content	Shoot Se content
Root biomass																	
Shoot biomass	0.904 * *																
Root/shoot ratio	0.246	−0.187															
Chlorophyll a content	0.967 * *	0.908 * *	0.170														
Chlorophyll b content	0.926 * *	0.809 * *	0.246	0.917 * *													
Chlorophyll a/b	0.480	0.556 *	−0.043	0.577 *	0.206												
Carotenoid content	0.948 * *	0.939 * *	0.035	0.956 * *	0.906 * *	0.491											
SOD activity	−0.469	−0.172	−0.636 *	−0.294	−0.488	0.272	−0.298										
POD activity	0.459	0.158	0.637 *	0.288	0.488	−0.282	0.320	−0.989 * *									
CAT activity	−0.627 *	−0.413	−0.435	−0.527 *	−0.683 * *	0.116	−0.479	0.804 * *	−0.762 * *								
Soluble protein content	−0.297	0.027	−0.670 * *	−0.142	−0.402	0.465	−0.151	0.959 * *	−0.967 * *	0.731 * *							
Proline content	0.734 * *	0.529 *	0.400	0.636 *	0.854 * *	−0.188	0.669 * *	−0.824 * *	0.831 * *	−0.854 * *	−0.792 * *						
MDA content	−0.406	−0.118	−0.589 *	−0.246	−0.511	0.442	−0.219	0.943 * *	−0.908 * *	0.844 * *	0.939 * *	−0.841 * *					
Relative conductivity	−0.633 *	−0.298	−0.729 * *	−0.519 *	−0.684 * *	0.115	−0.468	0.930 * *	−0.920 * *	0.885 * *	0.894 * *	−0.887 * *	0.90[illegible] * *				
TF	−0.854 * *	−0.960 * *	0.233	−0.897 * *	−0.850 * *	−0.443	−0.912 * *	0.080	−0.063	0.414	−0.071	−0.539 *	0.108	0.258			
Root Se content	0.795 * *	0.619 *	0.338	0.742 * *	0.937 * *	−0.097	0.745 * *	−0.673 * *	0.677 * *	−0.825 * *	−0.645 * *	0.968 * *	−0.742 * *	−0.817 * *	−0.677 * *		
Shoot Se content	0.265	−0.083	0.707 * *	0.147	0.447	−0.539 *	0.132	−0.864 * *	0.877 * *	−0.726 * *	−0.956 * *	0.793 * *	−0.913 * *	−0.874 * *	0.054	0.697 * *	

Note: * * = Correlation is significant at the 0.01 level (2-tailed). * = Correlation is significant at the 0.05 level (2-tailed). N = 15. TF = translocation factor.

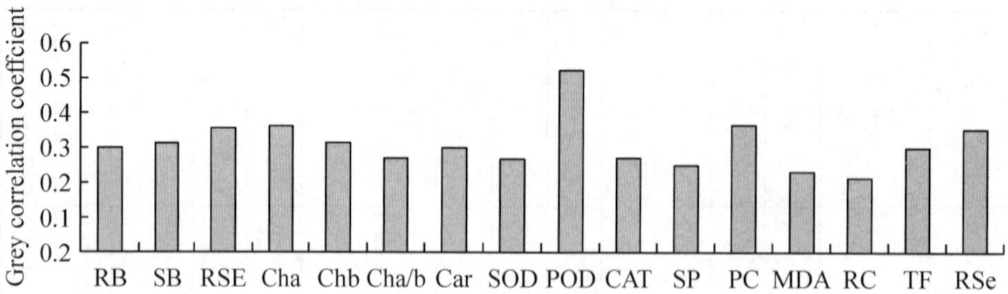

Note: RB = root biomass; SB = shoot biomass; RSR = root/ shoot ratio; Cha = chlorophyll a content; Chb = chlorophyll b content; Cha/b = chlorophyll a/b; Car = carotenoid content; POD = POD activity; SOD = SOD activity; CAT = CAT activity; SP = soluble protein content; PC = proline content; MDA = malonaldehyde content; RC = relative conductivity; TF = translocation factor; RSe = root Se content.

Figure 3-42 Grey correlation coefficients of the different parameters with the shoot Se content

4 Discussion

ABA is one of the most important hormones in plants and plays an important regulatory role in plant growth and development (Cao et al., 2017; Zhu et al., 2017). Exogenous ABA alleviates salt stress in tomato seedlings and promotes their growth (Hu et al., 2021). In Cd-contaminated soil, exogenous ABA promotes the growth of *S. photeinocarpum* and inhibites the growth of *B. pilosa* (Wang et al., 2016; Liu et al., 2017). The above findings show that ABA has different effects depending on plant species. In this study, only the 20 μmol/L ABA increased the biomass of *C. betacea* seedlings under Se stress, while other concentrations (1, 5, and 10 μmol/L) of ABA decreased the biomass to some extent. This result suggests that the response of plants to exogenous ABA might be genotype-dependent (Wang et al., 2016; Cao et al., 2017; Zhu et al., 2017; Hu et al., 2021), and the lower concentration of ABA (<20 μmol/L) inhibits and higher concentration of ABA (20 μmol/L) promotes the growth of *C. betacea*. The promoting effects of higher concentration of ABA on the growth of *C. betacea* seedlings need further study. In the present study, ABA increased the root/shoot ratio of *C. betacea* seedlings under Se stress, indicating that the inhibitory effect of ABA on the shoots of *C. betacea* seedlings was greater than that on the roots. Low concentrations of ABA (<20 μmol/L) inhibited the growth of *C. betacea* seedlings, while higher concentrations (to 20 μmol/L) promoted the growth. Notably, low concentrations of ABA have been shown to inhibit the growth, promote bud dormancy, and reduce plant growth rate (Hu et al., 2021).

Exogenous ABA treatment increased the content of chlorophyll a, chlorophyll b, and carotenoid in tomato seedlings under salt stress (Hu et al., 2021). In Cd-contaminated soil, ABA increased the chlorophyll content in *S. photeinocarpum*, but decreased chlorophyll content in *B. pilosa* under similar ABA treatment (Wang et al., 2016; Liu et al., 2017). Therefore, different plant species have various responses to ABA for the chlorophyll content, and ABA can mitigate the chlorophyll degradation of some plants under stress conditions (Sah et al., 2016; Xiang et al., 2021). In this study, only 20 μmol/L ABA increased the content of chlorophyll a, chlorophyll b, and carotenoid in *C. betacea* seedlings under Se stress, while the concentrations of 1, 5, and 10 μmol/L ABA decreased these parameters. In addition, different concentrations of ABA decreased the chlorophyll a/b under Se stress. These results indicate that the different plant species exhibit various responses to the same ABA concentration, which could be related to the ABA receptor ABAR/CHLH in the chloroplast of plant cells, which can catalyze cell chlorophyll synthesis under suitable ABA concentration (Shang, 2010). On the other hand, ABA alleviate the stress of Se to plants by improving the resistance of plants (Sah et al., 2016). Therefore, only 20 μmol/L ABA could promote the chlorophyll synthesis of *C. betacea* seedlings.

Metabolic processes that occur during the normal life cycle of plants are often accompanied by the production of reactive oxygen species (ROS). Under Se stress conditions, the plant cells accumulate ROS, which causes the oxidative stress and induces the programmed cell death (Quan et al., 2008; Kamran et al., 2019; Dai et al., 2020; Farooq et al., 2022). Plants have enzymatic and non-enzymatic antioxidant protection systems that facilitate ROS scavenging. The enzymatic antioxidants include SOD, CAT, and POD, whereas the non-enzymatic antioxidants include osmotic regulation substances, such as proline and soluble proteins (Liang et al., 2003). ABA at 50 μmol/L increased the POD, SOD, and CAT activities in tomato (Hu et al., 2021). Under low-temperature stress, exogenous ABA increased the content of proline and soluble protein in tomato seedlings and reduced the content of MDA and relative conductivity (Duan et al., 2020). In this study, ABA decreased the SOD activity of *C. betacea* seedlings, but increased the POD activity to some extent under Se stress. Only 20 μmol/L ABA decreased the CAT activity of *C. betacea* seedlings. These results contradict some previous studies (Duan et al., 2020; Hu et al., 2021), suggesting that ABA stressed *C. betacea* seedlings. This phenomenon is consistent with a previous study on *C. betacea* seedlings under different concentrations of Se stress (Lin et al., 2020). This can be attributed to tissue-organ specificity and developmental stage specificity of POD and CAT activities in plants. The activities of POD and CAT is the lowest at the seedling stage (Liang et al., 2003; Yang et al., 2012). Moreover, ABA can up-regulate the ROS-related genes, increase the glutathione content, and decrease the content of H_2O_2 and MDA to increase the resistance of plants to stress conditions (Yao et al., 2017). In this study, ABA increased the proline content in *C. betacea* seedlings, but decreased the soluble protein content, MDA content, and relative conductivity under Se stress. These results are contrary to previous studies (Duan et al., 2020; Hu et al., 2021), indicating that *C. betacea* seedlings suffered ABA stress. Exogenous ABA caused similar stress-related physiological changes in *C. betacea* seedlings under different concentrations of Se stress (Lin et al., 2020), which might be related to the effects of ABA on Se accumulation in *C. betacea* seedlings. In addition, exogenous ABA application causes endogenous ABA accumulation in plants (Sehar et al., 2022; Sharma et al., 2022), which drives the stomatal closure and regulates the expressions of osmoregulation related genes (such as *p5cs* gene), thereby reducing the transpiration rate and increasing the osmoregulatory substances content (Silva-Ortega et al., 2008).

Previous studies have reported that plant growth regulators such as diethyl aminoethyl hexanoate (DA-6) and salicylic acid can promote Se uptake in grapevine (Liu, 2021; Li et al., 2022b). In *C. betacea* seedlings, indoleacetic acid (IAA) at 60 ~ 90 mg/L promotes plant growth and Se accumulation at an optimal concentration of 60 mg/L (Huan et al., 2021). Similarly, biostimulants such as potassium fulvic acid and humic acids improve the Se accumulation in onion and rice (Zhang, 2020; Li et al., 2022a). In this study, only 10 and 20 μmol/L ABA increased the root Se content in *C. betacea* seedlings, while 1, 5, 10, and 20 μmol/L ABA increased the shoot Se content under Se stress. The root Se content had a parabolic relationship with the ABA concentration, while the shoot Se content had a linear function relationship with the ABA concentration. These results are consistent with the findings reported by Li et al. (2022b) on the effect of salicylic acid on Se accumulation in grapevine. Thus, ABA could promote Se uptake in *C. betacea* seedlings under Se stress, like ABA promotes Cd accumulation in the Cd-hyperaccumulators *S. photeinocarpum* and *B. pilosa* (Wang et al., 2016; Liu et al., 2017), and promote Se transport from roots to shoots when the concentration of ABA is lower than 20 μmol/L, but the associated mechanism and promoting effects of higher concentrations (>20 μmol/L) of ABA on the Se uptake of *C. betacea* seedlings need further study. The increased TF of *C. betacea* seedlings at 1, 5, and 10 μmol/L ABA also indicates that ABA could promote Se transport from roots to shoots. In addition, the root Se content was correlated with the biomass, photosynthetic pigment content, antioxidant enzyme activity, osmotic regulation substance content, MDA content, and relative conductivity. These results indicate that ABA promoted the Se uptake of *C. betacea* seedlings by improving its physiological resistance. Meanwhile, the shoot Se content was correlated with the root/shoot ratio, chlorophyll a/b, antioxidant enzyme activities, osmotic regulation substance content, MDA content, and relative conductivity. The grey relational analysis showed that the POD activity and proline content had the first two highest correlations with the shoot Se content. Therefore, the POD activity and proline content could contribute to ABA improving Se uptake in shoots by *C. betacea* seedlings. These results also indicate that the improvement of physiological resistance may be the reason for the increase in shoot Se content. Although our study just investigated the Se accumulation in *C. betacea* seedlings, the Se accumulation in fruits of *C. betacea* could be also increased by ABA according to the previous studies (Huang, 2021; Liu, 2021). The promoting Se uptake effects of ABA on fruits of *C. betacea* need to be further studied.

5 Conclusion

Under Se stress, ABA promotes the Se uptake in *C. betacea* seedlings, but only the concentration at 20 μmol/L promotes the growth of *C. betacea* seedlings. The shoot Se content is most closely correlated with the POD activity and proline content. The future work should assess the mechanism underlying ABA promoting effects on the Se uptake in *C. betacea* and replicate these findings in *C. betacea* fields.

References

[1] BOURSIAC Y, LERAN S, CORRATGE-FAILLIE C, et al. ABA transport and transporters[J]. Trends in Plant Science, 2013, 18(6):325-333.

[2] CAO C Y, CHEN X, WU Y. Advance in the regulation of abscisic acid on seed germination and dormancy[J]. Chinese Bulletin of Life Sciences, 2017, 29(6):582-588.

[3] CUTLER S R, RODRIGUEZ P L, FINKELSTEIN R R, et al. Abscisic acid: emergence of a core signaling network[J]. Annual Review of Plant Biology, 2010, 61(1):651-679.

[4] DAI Z, RIZWAN M, GAO F, et al. Nitric oxide alleviates selenium toxicity in rice by regulating antioxidation, selenium uptake, speciation and gene expression [J]. Environmental Pollution, 2020, 257:113540.

[5] DONG Q, LI S M, GAO S J, et al. Comparison and comprehensive analysis the quality of *Cyphomandra betacea* from different provenance [J]. Food and Fermentation Industries, 2022, 48(4):266-273.

[6] DUAN S Y, WANG M L, ZHANG X C, et al. Effects of exogenous abscisic acid, salicylic acid and gibberellin on chilling resistance of tomato seedlings[J]. Journal of Beijing University of Agriculture, 2020, 35(3):50-56.

[7] EL-RAMADY, H, ABDALLA, et al. Selenium in soils under climate change, implication for human health[J]. Environmental Chemistry Letters, 2015, 13(1):1-19.

[8] FAROOQ M U, ISHAAQ I, BARUTCULAR C, et al. Mitigation effects of selenium on accumulation of cadmium and morpho-physiological properties in rice varieties[J]. Plant Physiology and Biochemistry, 2022, 170:1-13.

[9] GUPTA K, WANI S H, RAZZAQ A, et al. Abscisic acid:role in fruit development and ripening[J]. Frontiers in Plant Science, 2022, 13:817500.

[10] HAO Z B, CANG J, XU Z. Plant physiology experiment[M]. Harbin Institute of Technology Press, Harbin, China, 2004.

[11] HOSSAIN A, SKALICKY M, BRESTIC M, et al. Selenium biofortification:roles, mechanisms, responses and prospects[J]. Molecules, 2021, 26(4):881.

[12] HU E, LIU M, ZHOU R, et al. Relationship between melatonin and abscisic acid in response to salt stress of tomato[J]. Scientia Horticulturae, 2021, 285:110176.

[13] HUAN Y, YANG L, LIU Q, et al. Effects of indole acetic acid on the growth and selenium absorption characteristics of *Cyphomandra betacea* seedlings[J]. Acta Physiologiae Plantarum, 2021, 43(5):74.

[14] HUANG K. Study on the accumulation characteristics of selenium in grapevine and its regulation mechanism by *Paraglomus occultum* [D]. Sichuan Agricultural University, Chengdu, China, 2021.

[15] JIA X T, ZHAO Q Y, Zhang J M, et al. Research progress on selenium-enriched livestock products[J]. Food and Nutrition in China, 2021, 27(2):26-31, 10.

[16] JIANG Y, ZENG Z H, YANG Q S, et al. Selenium (Se) uptake and transformation mechanisms and physiological function in plant: A review [J]. Chinese Journal of Applied

Ecology, 2016, 27(12):4067-4076.

[17]KAMRAN M, PARVEEN A, AHMAR S, et al. An overview of hazardous impacts of soil salinity in crops, tolerance mechanisms, and amelioration through selenium supplementation[J]. International Journal of Molecular Sciences, 2019, 21(1):148.

[18]KARIMI R, EBRAHIMI M, AMERIAN M. Abscisic acid mitigates NaCl toxicity in grapevine by influencing phytochemical compounds and mineral nutrients in leaves[J]. Scientia Horticulturae, 2021, 288:110336.

[19]KAUR G, KALIA A, SODHI H S. Selenium biofortification of Pleurotus species and its effect on yield, phytochemical profiles, and protein chemistry of fruiting bodies[J]. Journal of Food Biochemistry, 2018, 42(2):e12467.

[20]KIKKERT J, BERKELAAR E. Plant uptake and translocation of inorganic and organic forms of selenium[J]. Archives of Environmental Contamination and Toxicology, 2013, 65(3):458-465.

[21]KUROMORI T, SEO M, SHINOZAKI K. ABA transport and plant water stress responses[J]. Trends in Plant Science, 2018, 23(6):513-522.

[22]LI F Y, LU W H, ZHANG N M, et al. Effects of different fertilization treatments on selenium absorption, transportation and yield of onion[J]. Vegetables, 2022a, 41(1):24-29.

[23]LI Z Y, FAN R, PENG X M, et al. Salicylic acid alleviates selenium stress and promotes selenium uptake of grapevine[J]. Physiology and Molecular Biology of Plants, 2022b, 28(3):625-635.

[24]LIANG Y R, HU X R, ZHANG Y L, et al. Progress on physiological function research of plant peroxidase[J]. Journal of Inner Mongolia Agricultural University, 2003, 24(2):110-113.

[25] LIN L, SUN J, CUI T, et al. Selenium accumulation characteristics of *Cyphomandra betacea* (*Solanum betaceum*) seedlings[J]. Physiology and Molecular Biology of Plants, 2020, 26(7):1375-1383.

[26] LIU L. Physiological and molecular mechanisms of plant growth regulators promoting selenium enrichment in grapes[D]. Sichuan Agricultural University, Chengdu, China, 2021.

[27]LIU L, MA Q, LIN L, et al. Effects of exogenous abscisic acid on cadmium accumulation in two ecotypes of hyperaccumulator *Bidens pilosa*[J]. Environmental Progress

& Sustainable Energy, 2017, 36(6):1643-1649.

[28] LIU L, WANG T, SUI LY, et al. The selenium accumulation characteristics of grape seedlings[J]. IOP Conference Series: Earth and Environmental Science, 2019, 330(4):42042.

[29] LIU D H, ZHOU W B, ZHU D W, et al. General survey on physiological function of selenium in plant[J]. Journal of Mountain Agriculture and Biology, 2005, 24(3):253-259.

[30] LIU Q, ZHANG H, WANG Y, et al. Effect of straw biochar from different crops on growth and nutrient uptake by peach (*Prunus davidiana*) seedlings[J]. Acta Physiologiae Plantarum, 2022, 44(10):104.

[31] LUO H Z. Effects of mutual grafting on the growth and selenium enrichment of cherry tomato progeny of two genotypes[D]. Sichuan Agricultural University, Chengdu, China, 2019.

[32] PAN S, LU R, LI H, et al. Mutual intercropping affects selenium uptake of eggplant seedlings[J]. International Journal of Environmental Analytical Chemistry, 2021, 101(15):2866-2875.

[33] PELEG Z, BLUMWALD E. Hormone balance and abiotic stress tolerance in crop plants[J]. Current Opinion in Plant Biology, 2011, 14(3):290-295.

[34] PU Y. Effects of piriformis indica on the biomass, quality and selenium accumulation of Chinese cabbage[D]. Yangtze University, Jingzhou, China. 2021.

[35] PUCCINELLI M, MALORGIO F, PEZZAROSSA, B. Selenium enrichment of horticultural crops[J]. Molecules, 2017, 22(6):933.

[36] QUAN L J, ZHANG B, SHI W W, et al. Hydrogen peroxide in plants: a versatile molecule of the reactive oxygen species network[J]. Journal of Integrative Plant Biology, 2008, 50(1):2-18.

[37] RADAWIEC A, SZULC W, RUTKOWSKA B. Selenium biofortification of wheat as a strategy to improve human nutrition[J]. Agriculture, 2021, 11(2):144.

[38] RAGHAVENDRA A S, GONUGUNTA V K, CHRISTMANN A, et al. ABA perception and signaling[J]. Trends in Plant Science, 2010, 15(7):395-401.

[39] RAMASAMY S, NANDAGOPAL J G T, BALASUBRAMANIAN M, et al. Effect of abscisic acid and selenium foliar sprays on drought mitigation in tomato (*Solanum lycopersicum* L.)[J]. Materials Today: Proceedings, 2022, 48:191-195.

[40]RASTMANESH F, MOORE F, KESHAVARZI B. Speciation and phytoavailability of heavy metals in contaminated soils in Sarcheshmeh area, Kerman Province, Iran[J]. Bulletin of Environmental Contamination and Toxicology, 2010, 85(5):515-519.

[41]SAH S K, REDDY K R, LI J. Abscisic acid and abiotic stress tolerance in crop plants[J]. Frontiers in Plant Science, 2016, 7(571):571.

[42] SEHAR Z, IQBAL N, FATMA M, et al. Ethylene suppresses abscisic acid, modulates antioxidant system to counteract arsenic-inhibited photosynthetic performance in the presence of selenium in mustard[J]. Frontiers in Plant Science, 2022, 13:852704.

[43] SHANG Y. An abscisic acid receptor ABAR/CHLH mediated ABA signaling pathway in *Arabidopsis*[D]. Tsinghua University, Beijing, China, 2010.

[44] SHARMA A, RAMAKRISHNAN M, KHANNA K, et al. Brassinosteroids and metalloids: regulation of plant biology [J]. Journal of Hazardous Materials, 2022, 424:127518.

[45]SHEN G M, NIU J K, DENG Z X. Abscisic acid treatment alleviates cadmium toxicity in purple flowering stalk (*Brassica campestris* L. ssp. *chinensis* var. *purpurea* Hort.) seedlings[J]. Plant Physiology and Biochemistry, 2017, 118:471-478.

[46]SHI H P. Selenium and its physiological roles in plants[J]. Chinese Bulletin of Botany, 1995, 12(A1):13-36.

[47]SILVA-ORTEGA C O, OCHOA-ALFARO A E, REYES-AGüERO J A, et al. Salt stress increases the expression of *p5cs* gene and induces proline accumulation in cactus pear [J]. Plant Physiology and Biochemistry, 2008, 46(1):82-92.

[48]SUPRIATIN S, WENG L, COMANS, et al. Selenium speciation and extractability in Dutch agricultural soils[J]. Science of the Total Environment, 2015, 532, 368-382.

[49]WANG J, LIN L, LUO L, et al. The effects of abscisic acid (ABA) addition on cadmium accumulation of two ecotypes of *Solanum photeinocarpum* [J]. Environmental Monitoring and Assessment, 2016, 188(3):182.

[50]WANG X L. Application of grey relation analysis theory to choose high reliability of the network node[J]. Journal of Physics:Conference Series, 2019, 1237(3):032056.

[51]XIANG H T, ZHENG D F, HE N, et al. Research progress on the physiological response of plants to low temperature and the amelioration effectiveness of exogenous ABA [J]. Acta Prataculturae Sinica, 2021, 30(1):208-219.

[52]XU X, FANG L, LI L, et al. Abscisic acid inhibits a symbiotic germination of

immature seeds of *Paphiopedilum armeniacum* [J]. International Journal of Molecular Sciences, 2020, 21(24):9561.

[53] YANG D Y, ZHANG L Z, XU S Q. Study on catalase activity in watermelon during its different growth periods [J]. Journal of Anhui Agricultural Science, 2012, 40(18):9604-9606.

[54] YAO X M, JI J, YUE J Y, et al. Reactive oxygen metabolism and related gene expression in *Platycladus orientalis* under salt stress[J]. Acta Botanica Boreali-Occidentalia Sinica, 2017, 37(1):105-114.

[55] ZAFEIRIOU I, GASPARATOS D, IOANNOU D, et al. Selenium biofortification of lettuce plants (*Lactuca sativa* L.) as affected by Se species, Se rate, and a biochar co-application in a calcareous soil[J]. Agronomy, 2022, 12(1):131.

[56] ZHANG C M. The regulatory effects of selenium nutrition in soil-rice system and cost-effectiveness analysis[D]. Southwest University, Chongqing, China, 2018.

[57] ZHANG H Y. Effect of humic acid on selenium and cadmium uptake in rice (*Oryza sativa* L.)[D]. China University of Geosciences, Wuhan, China, 2020.

[58] ZHANG J C, WANG X F, WANG X N, et al. Abscisic acid alleviates iron deficiency by regulating iron distribution in roots and shoots of apple [J]. Scientia Horticulturae, 2020, 262:109018.

[59] ZHANG Q L, ZHANG F X, FENG Z X, et al. The preliminary experimental study on introduction and cultivation of tree tomato [J]. Shandong Forestry Science and Technology, 2005, 3:25-26.

[60] ZHANG R, LIU Q, XU X, et al. An amino acid fertilizer improves the emergent accumulator plant *Nasturtium officinale* R. Br. phytoremediation capability for cadmium-contaminated paddy soils[J]. Frontiers in Plant Science, 2022, 13:1003743.

[61] ZHOU F, PENG Q, WANG M, et al. Influence of processing methods and exogenous selenium species on the content and in vitro bioaccessibility of selenium in *Pleurotus eryngii*[J]. Food Chemistry, 2021, 338:127661.

[62] ZHU X C, ZHANG H M, NAN W B. Research progress on regulation of ABA in plant root development[J]. Plant Physiology Journal, 2017, 53(7):1123-1130.

Chapter Ⅳ Grafting

Section Ⅰ *Solanum* rootstocks

1 Introduction

Selenium (Se) is an essential trace element for human health, playing important roles in the central nervous system, endocrine system, cardiovascular system, and immune system (Roman et al., 2014). However, rapid urbanization and industrialization have led to soil contamination of Se in several regions worldwide, such as the Southwestern United States, Torun in Poland, and Enshi in China, primarily due to electronic waste and mining activities (Zhang et al., 2023). Soil Se contamination inhibits the growth of crops, often leading to plant death and ultimately limiting agricultural production (Lanza et al., 2021). Moreover, excessive Se induces oxidative and nitroxidative stress in crops, affecting their metabolism (Gupta et al., 2017; Józwiak et al., 2019). Therefore, remediation of Se-contaminated soil becomes imperative to mitigate these negative effects on sustainable agriculture.

Grafting is a common asexual propagation technique for horticultural crops which has been widely used in recent years for variety selection, quality improvement, and resistance enhancement, resulting in lucrative economic benefits (Maurya et al., 2019). Grafting can improve crop growth and yield; for example, in tomato, grafting leads to increased stem thickness, leaf number, fruit yield per plant, and fruit weight per plant (Guo, 2018). Grafting could also increase the photosynthetic pigment content, enhance the ability to capture low light, and increase the photosynthetic rate while significantly upregulating the photosynthesis-related genes in grafted tomato plants (Ren et al., 2019). Furthermore, grafting can increase the antioxidant enzyme activity and the content of osmoregulatory substances in plant leaves and roots, thus improving salt stress tolerance (Shahid et al.,

2019). In addition, as a root replacement cultivation technique, grafting has a greater impact on the uptake of mineral nutrients in crops (Khankahdani et al., 2019). Grafting citrus onto the rootstock of Volkamer lemon and grafting tomato on the rootstock of wild tomato increase the uptake of trace elements such as iron, zinc, and manganese in scions (Chen et al., 2019; Khalid et al., 2022). Grafting *Nasturtium officinale* with broccoli rootstock promotes Se accumulation of *N. officinale* cuttings (Huang et al., 2020). Similarly, wild tomato rootstock increases Se uptake in eggplant seedlings (Lu et al., 2021). These studies indicate that grafting can promote Se uptake in vegetable crops; however, research on grafting fruit trees for Se uptake remains limited.

Cyphomandra betacea is a fruit tree belonging to Solanaceae family, and its fruits are rich in a variety of minerals, pectin, and vitamin C, contributing to its high ornamental and nutritional value (Ordoñez-Santos et al., 2020). A previous study showed that the Se accumulation capacity of *C. betacea* was lower than that of common vegetable crops (Lin et al., 2020a). Implementing measures to improve Se uptake by *C. betacea* could thus improve the commercial value of its fruits. Tomato (*Solanum lycopersicum*), eggplant (*Solanum melongena*), and nightshade (*Solanum nigrum*) belong to the Solanaceae family, and studies have shown that tomato rootstock can increase the uptake of mineral elements in nightshades, eggplants and tomatoes with the order of effectiveness being nightshade > eggplant > tomato (Wang et al., 2019). Tomato and eggplant rootstocks can also affect the Se uptake characteristics of vegetable crops (Luo, 2019; Ordoñez-Santos et al., 2020; Zhang, 2021). Tomato, eggplant, and nightshade rootstocks can also change the Se uptake characteristics of *C. betacea*. Therefore, in this experiment, we grafted *C. betacea* seedlings with tomato, eggplant, and nightshade rootstocks and evaluated the effects of their impact on the growth and Se accumulation of *C. betacea* seedlings. This study aimed to screen the most effective rootstock that could promote the growth and Se accumulation of *C. betacea*, providing valuable insights for Se-enriched *C. betacea* production.

2 Materials and methods

2.1 Materials

C. betacea seeds were collected from a seven-year-old fruiting tree at the Chengdu Campus of Sichuan Agricultural University (30°42′N, 103°51′E) in September 2021. Tomato and eggplant seeds were purchased from a company in Wenjiang District, Chengdu City,

China. Nightshade seeds and soil samples were collected from farmland around the Chengdu Campus of Sichuan Agricultural University. The soil was fluvo-aquic, and its basic physical and chemical properties were as follows: pH 7.58, total nitrogen (N) content of 1.44 g/kg, total phosphorus (P) content of 0.43 g/kg, total potassium (K) content of 24.55 g/kg, alkaline hydrolyzed N content of 58.68 mg/kg, available P content of 19.23 mg/kg, available K content of 58.33 mg/kg, and total Se content of 0.35 mg/kg.

2.2 Experimental design

The soil samples were treated in February 2022, air-dried, and passed through 5-mm mesh as described by Lin et al. (2020b). Each plastic pot (20 cm height, 21 cm diameter) contained 3.0 kg of air-dried soil samples. Na_2SeO_3 was added to the soil to make the final soil Se concentration of 5 mg/kg (Lin et al., 2022). The soil moisture content was maintained at 80% of field capacity for one month.

In February 2022, the seeds of *C. betacea*, tomato, eggplant, and nightshade were sown, and grafting occurred when plant seedlings reached a height of 10 cm. The scions of *C. betacea* seedlings were grafted onto the rootstocks of tomato, eggplant, and nightshade seedlings. The grafting method and post-grafting management of plants were conducted as described by Lin et al. (2016). In March 2022 (10 days after grafting), the surviving grafted seedlings were transplanted into pots containing Se-contaminated soil with each pot containing three plants. The treatments included un-grafted (UG), tomato rootstock grafting (TG), eggplant rootstock grafting (EG), and nightshade rootstock grafting (SG). Each treatment was conducted in triplicate, and the soil moisture content was maintained at 80% of the field capacity. After 40 days of growth in Se-contaminated soils, the plants were harvested.

2.3 Determination of biomass

The plants were divided into rootstock root, rootstock stem, and scion. Each part was washed with tap water followed by deionized water, and then dried at 80 ℃ until constant weight was achieved to determine their dry biomass (Li et al., 2022). For the UG plants, the stem was collected, and the lower part of the stem (with the same height as the grafted plants) was treated as the rootstock, while the upper stem with the leaves was treated as the scion. Thus, the rootstock root, rootstock stem, and scion biomass were determined using an electronic balance.

2.4 Determination of photosynthetic characteristic parameters

Before harvested, the fifth mature leaves from the top of the plants were sampled for measuring the photosynthetic parameters, including the net photosynthetic rate (Pn), transpiration rate (Tr), stomatal conductance (Gs), and intercellular CO_2 concentration (Ci), using a Li-6400 photosynthetic system (LI-COR, Lincoln, NE, USA). The determinations were conducted between 9:30 and 11:30 am under a light intensity of 1,000 μmol/(m^2/s) and a CO_2 concentration of 400 μmol/mol (Li et al., 2022).

2.5 Determination of photosynthetic pigment content

Before harvested, the fifth mature leaves from the top of the plants were also used to measure the content of photosynthetic pigments (chlorophyll a, chlorophyll b, total chlorophyll and carotenoid). Photosynthetic pigment content was extracted using the acetone-ethanol extraction method and determined at 663, 645, and 470 nm wavelengths using a spectrophotometer (Summit, Shanghai, China) according to the report of Hao et al. (2004).

2.6 Determination of antioxidant enzyme activity and soluble protein content

Before harvested, the sixth mature leaf from the top of the plants was used to measure the antioxidant enzyme [superoxide dismutase (SOD), peroxidase (POD) and catalase (CAT)] activities and soluble protein content. The leaf samples were homogenized in 6 mL extraction buffer (0.05 M potassium phosphate buffer [pH 7.0]) to extract the enzyme at 4 ℃. After centrifugation, the supernatants were determined using the methods of nitroblue tetrazolium photoreduction, guaiacol colorimetric, UV spectrophotometry and Coomassie brilliant blue staining for SOD activity, POD activity, CAT activity and soluble protein content, respectively according to the methods by Lin et al. (2020b) and Hao et al. (2004).

2.7 Determination of Se content

The dried samples were grounded and sieved through a 0.15-mm mesh nylon sieve. Then, the ground samples were digested with a 4:1 (v/v) nitrate-perchloric acid and reduced with hydrochloric acid. The solutions were used to determine the Se content. The Se content in rootstock roots, rootstock stems and scions were determined using a hydride generation-atomic fluorescence spectrometry (AFS-9700, Beijing Haiguang Instrument Co., Ltd.,

Beijing, China) (Li et al., 2022). The bioconcentration factor (BCF) was calculated as Se content in plants/soil, and the translocation factor (TF) was calculated as Se content in the upper organ/lower organ (Rastmanesh et al., 2010).

2.8 Statistical analyses

Statistical analyses were conducted using SPSS 20.0 software (IBM, Chicago, IL, USA). Data was normalized and subjected to a homogeneity test before a one-way analysis of variance and Duncan's multiple range test ($P<0.05$). Data of each indicator was subjected to the principal component analysis (PCA) and cluster analysis, as described by Lin et al. (2023).

3 Results

3.1 Biomass

The rootstock root biomassof TG and SG *C. betacea* seedlings had no significant differences compared with UG, while the rootstock root biomass of EG was lower than that of UG (Figure 4-1). Moreover, the rootstock stem biomass of TG and SG *C. betacea* seedlings was higher than that of UG, while that of EG had no significant difference compared with UG (Figure 4-2). The scion biomass of TG, EG and SG *C. betacea* seedlings were higher than UG (Figure 4-3), with an order of TG > SG > EG > UG. The scion biomass of TG, EG and SG *C. betacea* seedlings increased by 146.1%, 23.2% and 94.5%, respectively, compared with UG.

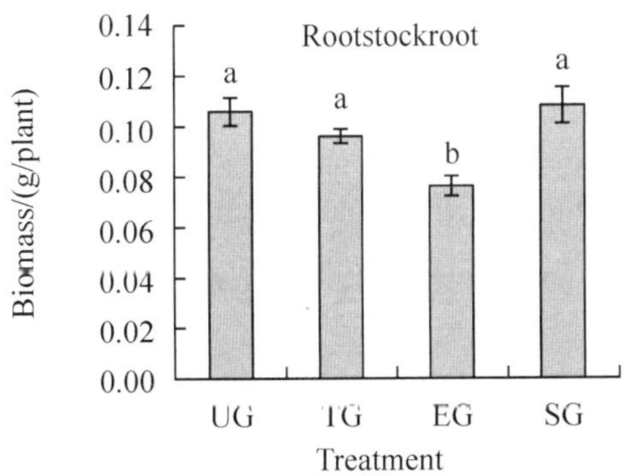

Note: Different lowercase letters indicate significant differences among the treatments (Duncan's multiple range test, $P < 0.05$).

Figure 4-1 Rootstock root biomass of *C. betacea* seedlings

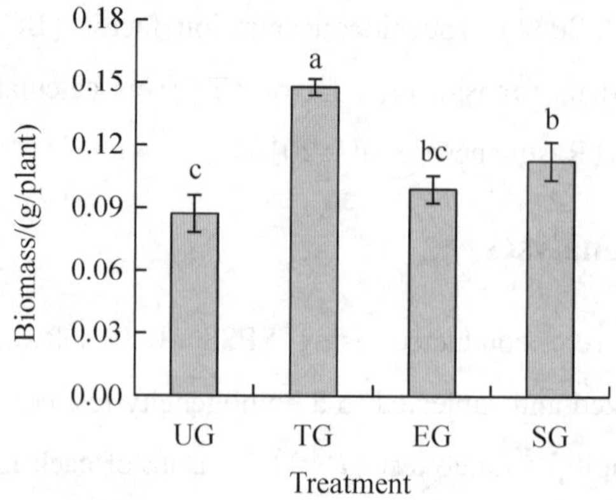

Note: Different lowercase letters indicate significant differences among the treatments (Duncan's multiple range test, $P < 0.05$).

Figure 4-2 Rootstock stem biomass of *C. betacea* seedlings

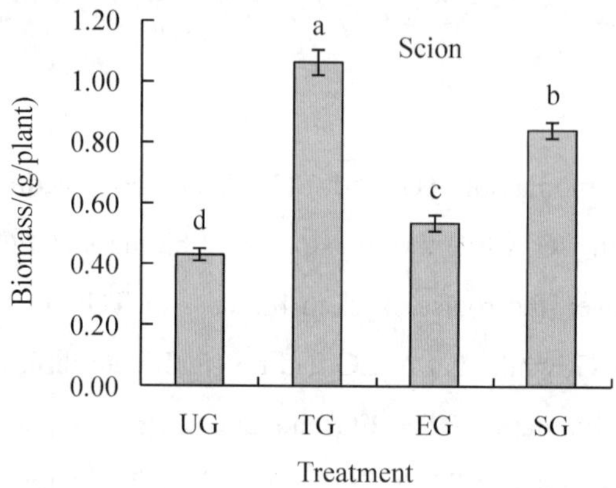

Note: Different lowercase letters indicate significant differences among the treatments (Duncan's multiple range test, $P < 0.05$).

Figure 4-3 Scion biomass of *C. betacea* seedlings

3.2 Photosynthetic pigment content

TG and SG increased chlorophyll a, chlorophyll b, and total chlorophyll content in *C. betacea* seedlings, while EG had no significant difference compared with UG (Table 4-1). Compared with UG, TG increased the content of chlorophyll a, chlorophyll b, and total chlorophyll by 26.4%, 27.2% and 26.6%, respectively. Similarly, SG increased the content of chlorophyll a, chlorophyll b, and total chlorophyll by 7.4%, 23.7% and 11.0%, respectively. Furthermore, TG increased the carotenoid content of *C. betacea* seedlings by 23.7%, while EG and SG had no significant differences compared with UG.

Table 4-1 Photosynthetic pigment content in *C. betacea* seedlings

Treatment	Chlorophyll a/ (mg/g)	Chlorophyll b/ (mg/g)	Totalchlorophyll/ (mg/g)	Carotenoid/ (mg/g)
UG	1. 170±0. 027c	0. 342±0. 012b	1. 512±0. 039c	0. 257±0. 004b
TG	1. 479±0. 038a	0. 435±0. 011a	1. 914±0. 049a	0. 318±0. 005a
EG	1. 225±0. 018bc	0. 363±0. 007b	1. 588±0. 025bc	0. 275±0. 007b
SG	1. 256±0. 011b	0. 423±0. 014a	1. 679±0. 025b	0. 270±0. 009b

Note: Different lowercase letters indicate significant differences among the treatments (Duncan's multiple range test, $P < 0.05$).

3. 3 Photosynthetic characteristic parameters

The Pn, Gs, and Tr of TG, EG, and SG of *C. betacea* seedlings were higher than UG, and TG had the highest Pn, Gs and Tr values (Table 4-2). Compared with UG, TG increased the Pn, Gs, and Tr by 38. 9%, 76. 7% and 43. 4%, respectively. TG and SG increased the Ci of *C. betacea* seedlings by 17. 1% and 12. 2%, respectively, while EG had no significant difference compared with UG.

Table 4-2 Photosynthetic characteristic parameters of *C. betacea* seedlings

Treatment	Pn/ [μmol CO_2/(m^2/s)]	Gs/ [mol H_2O/(m^2/s)]	Ci/ (μmol CO_2/mol)	Tr/ [mmol H_2O/(m^2/s)]
UG	3. 706±0. 063d	0. 0345±0. 0019d	188. 35±2. 68c	0. 730±0. 018c
TG	5. 146±0. 080a	0. 0610±0. 0010a	220. 62±6. 61a	1. 047±0. 023a
EG	4. 332±0. 093c	0. 0438±0. 0004c	196. 36±5. 87bc	0. 806±0. 013c
SG	4. 857±0. 120b	0. 0531±0. 0016b	211. 42±8. 92ab	0. 932±0. 049b

Note: Different lowercase letters indicate significant differences among the treatments (Duncan's multiple range test, $P < 0.05$).

3. 4 Antioxidant enzyme activity and soluble protein content

Compared with UG, TG and SG increased the POD activity of *C. betacea* seedlings by 70. 1% and 15. 3%, respectively, while EG had no significant difference (Table 4-3). The SOD activities of TG, EG, and SG in *C. betacea* seedlings were higher than the UG ones. Compared with UG, TG, EG, SG increased the SOD activity by 39. 7%, 24. 1% and 26. 7%, respectively. However, TG, EG, and SG decreased the CAT activity and soluble protein content of *C. betacea* seedlings.

Table 4-3 Antioxidant enzyme activity and soluble protein content in *C. betacea* seedlings

Treatment	POD activity/ [U/(g/min)]	SOD activity/ (U/g)	CAT activity/ [mg/(g/min)]	Soluble protein/ (mg/g)
UG	5,602±78c	192. 37±2. 40c	6. 333±0. 176a	27. 62±0. 67a
TG	9,530±211a	268. 69±4. 92a	4. 156±0. 221c	17. 61±0. 58d
EG	5,959±142c	238. 68±9. 46b	5. 535±0. 179b	21. 91±0. 82b
SG	6,459±75b	243. 69±9. 36b	4. 980±0. 244b	19. 81±0. 87c

Note: Different lowercase letters indicate significant differences among the treatments (Duncan's multiple range test, $P < 0.05$).

3.5 Se content, BCF and TF

The Se content in the rootstock root, rootstock stem, and scion of*C. betacea* seedlings exhibited the order of rootstock root > rootstock stem > scion (Figure 4-4, Figure 4-5, and Figure 4-6). The Se content of the rootstock roots of TG, EG, and SG of *C. betacea* seedlings was lower than the UG ones (Figure 4-4). Compared with UG, TG, EG, SG decreased the Se content in the rootstock roots by 59. 1%, 51. 6% and 33. 3%, respectively. The Se content in rootstock stems of *C. betacea* seedlings ranked as UG > SG > EG > TG (Figure 4-5). Compared with UG, TG, EG, SG decreased the Se content in rootstock stems by 32. 5%, 27. 0% and 15. 1%, respectively. Moreover, the Se content in scions of *C. betacea* seedlings ranked as SG > UG > EG > TG (Figure 4-6), and TG increased the Se content in scions by 6. 0% compared with UG.

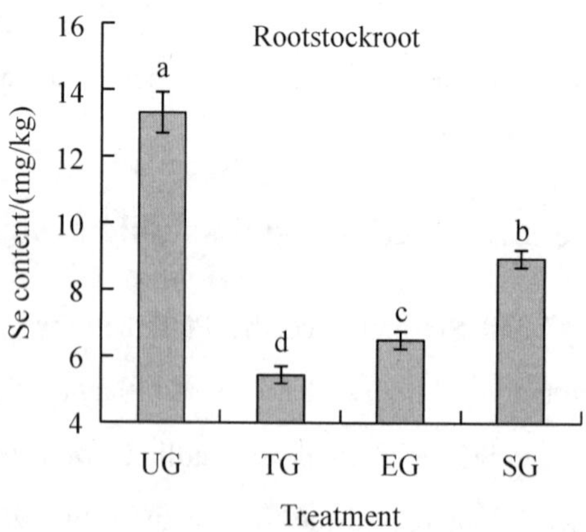

Note: Different lowercase letters indicate significant differences among the treatments (Duncan's multiple range test, $P < 0.05$).

Figure 4-4 Rootstock root Se content in *C. betacea* seedlings

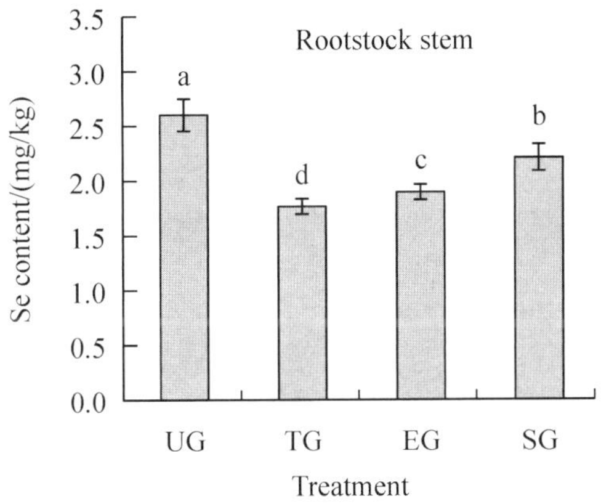

Note: Different lowercase letters indicate significant differences among the treatments (Duncan's multiple range test, $P < 0.05$).

Figure 4-5 Rootstock stem Se content in *C. betacea* seedlings

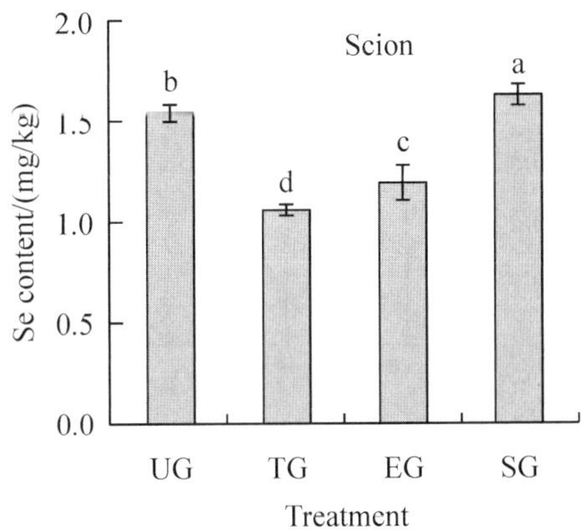

Note: Different lowercase letters indicate significant differences among the treatments (Duncan's multiple range test, $P < 0.05$).

Figure 4-6 Scion Se content in *C. betacea* seedlings

The rootstock root and rootstock stem BCFs of *C. betacea* seedlings exhibited the order of UG > SG > EG > TG (Figure 4-7 and Figure 4-8), while the scion BCF ranked as SG > UG > EG > TG (Figure 4-9). The $TF_{shoot/rootstock\ root}$ of TG, EG, and SG *C. betacea* seedlings was higher than UG (Figure 4-10); however, there as no significant difference among the grafting treatments. The $TF_{rootstock\ stem/rootstock\ root}$ of *C. betacea* seedlings ranked as TG > EG > SG > UG (Figure 4-11). SG increased the $TF_{scion/rootstock\ stem}$ of *C. betacea* seedlings, while TG and EG had no significant difference compared with UG (Figure 4-12).

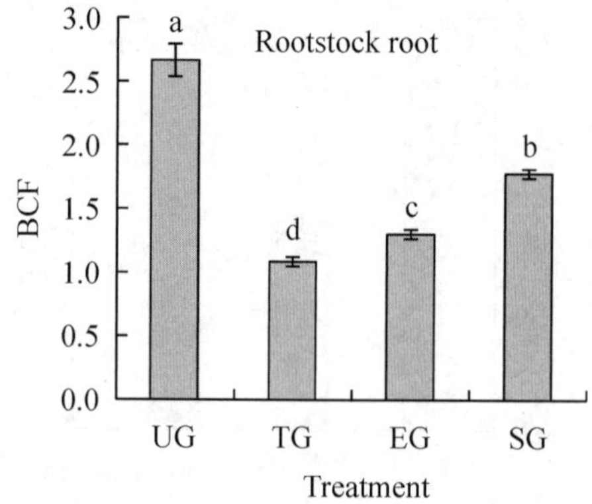

Note: Different lowercase letters indicate significant differences among the treatments (Duncan's multiple range test, $P < 0.05$).

Figure 4-7 Rootstock root bioconcentration factor (BCF) of *C. betacea* seedlings

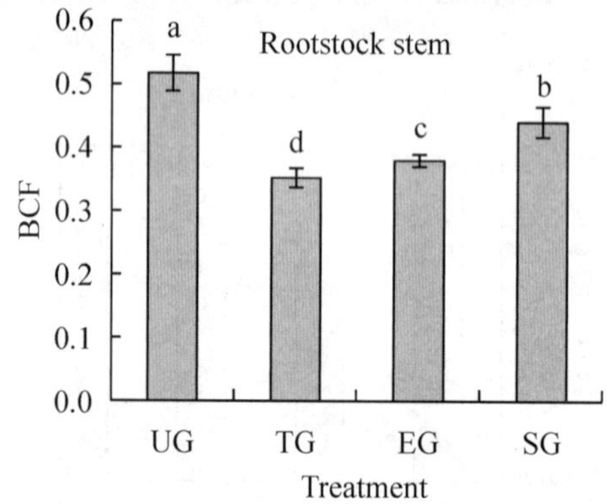

Note: Different lowercase letters indicate significant differences among the treatments (Duncan's multiple range test, $P < 0.05$).

Figure 4-8 Rootstock stem bioconcentration factor (BCF) of *C. betacea* seedlings

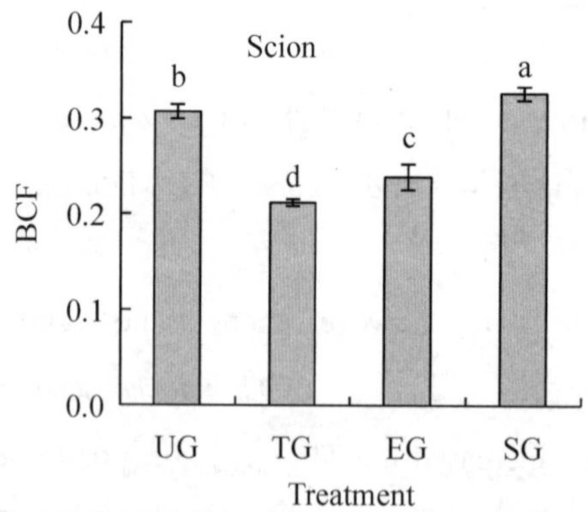

Note: Different lowercase letters indicate significant differences among the treatments (Duncan's multiple range test, $P < 0.05$).

Figure 4-9 Scion bioconcentration factor (BCF) of *C. betacea* seedlings

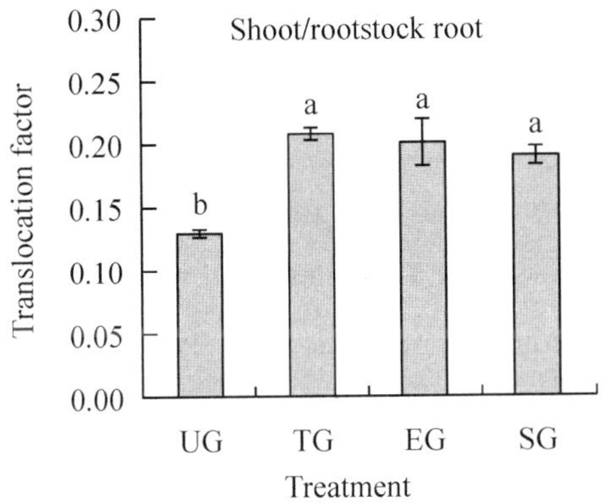

Note: Different lowercase letters indicate significant differences among the treatments (Duncan's multiple range test, $P < 0.05$).

Figure 4-10 Translocation factor (TF)$_{shoot/rootstock\ root}$ of *C. betacea* seedlings

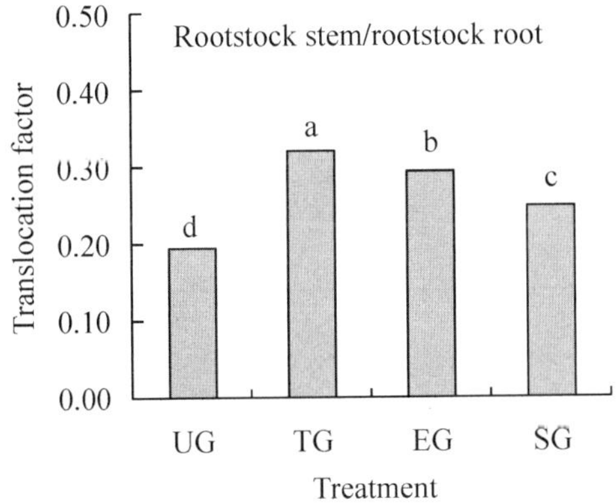

Note: Different lowercase letters indicate significant differences among the treatments (Duncan's multiple range test, $P < 0.05$).

Figure 4-11 Translocation factor (TF)$_{rootstock\ stem/rootstock\ root}$ of *C. betacea* seedlings

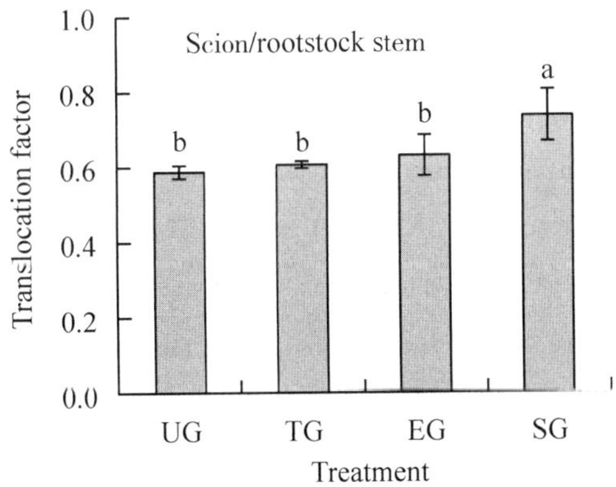

Note: Different lowercase letters indicate significant differences among the treatments (Duncan's multiple range test, $P < 0.05$).

Figure 4-12 Translocation factor (TF)$_{scion/rootstock\ stem}$ of *C. betacea* seedlings

3.6 Principal component analysis (PCA) and cluster analysis

PCA revealed that PC1 and PC2 accounted for 80.04% and 13.23% variance (total of 93.27%), respectively (Figure 4-13). The root biomass, scion Se content, CAT activity, soluble protein content, rootstock root Se content, and rootstock stem Se content were closely related. Similarly, the other indicators were also closely related. The cluster analysis grouped the indicators into two categories (Figure 4-14). The first cluster category comprised the root biomass, scion Se content, CAT activity, soluble protein content, rootstock root Se content, and rootstock stem Se content, and the second category comprised the other indicators. Thus, the PCA and cluster analysis had the same categories.

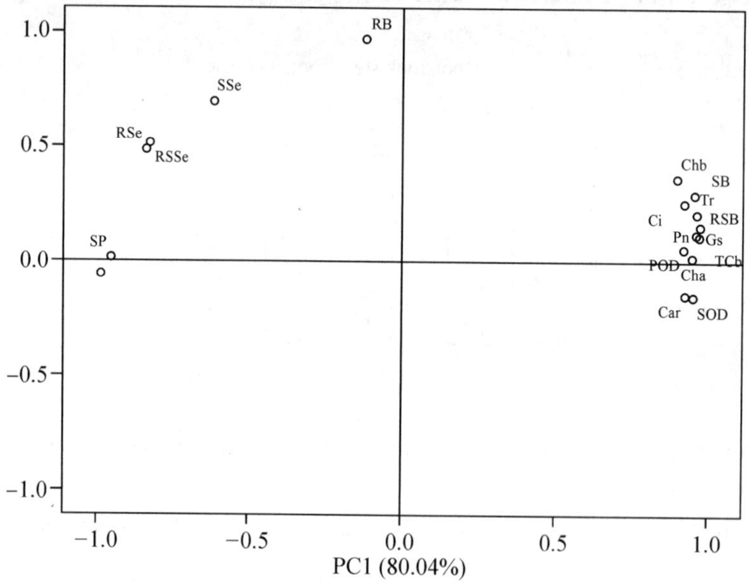

Note: RB = root biomass; RSB = rootstock stem biomass; SB = scion biomass; Cha = chlorophyll a content; Chb = chlorophyll b content; TCh = total chlorophyll content; Car = carotenoid content; Pn = net photosynthetic rate; Gs = stomatal conductance; Ci = intercellular CO_2 concentration; Tr = transpiration rate; POD = POD activity; SOD = SOD activity; CAT = CAT activity; SP = soluble protein content; RSe = rootstock root Se content; RSSe = rootstock stem Se content; SSe = scion Se content.

Figure 4-13 Principal component analysis (PCA)

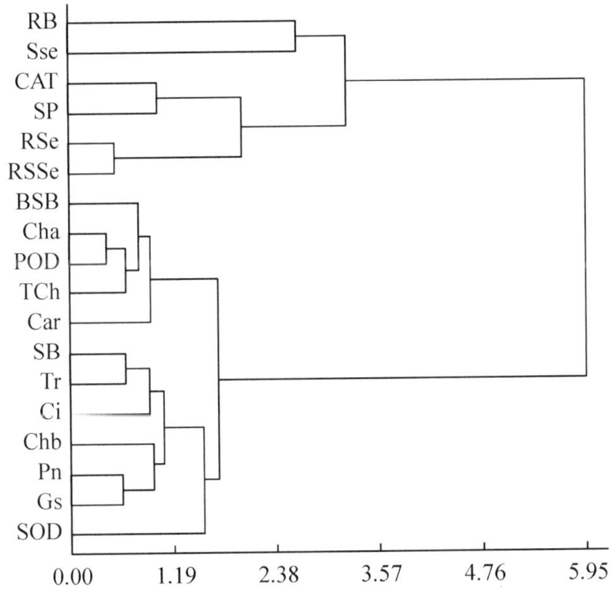

Note: RB = root biomass; RSB = rootstock stem biomass; SB = scion biomass; Cha = chlorophyll a content; Chb = chlorophyll b content, TCh = total chlorophyll content; Car = carotenoid content; Pn = net photosynthetic rate; Gs = stomatal conductance; Ci = intercellular CO_2 concentration; Tr = transpiration rate; POD = POD activity; SOD = SOD activity; CAT = CAT activity; SP = soluble protein content; RSe = rootstock root Se content; RSSe = rootstock stem Se content; SSe = scion Se content.

Figure 4-14 Cluster analysis

4 Discussion

Although the rootstock and scion transport and exchange substances through the xylem and phloem, the uptake of mineral element and transport of assimilates in plants are usually altered after grafting, indicating that different rootstocks have different effects on the growth of the scion (He et al., 2017; Nogueira et al., 2021). Some studies have shown that grafting can increase the biomass of eggplant (Lu et al., 2021), watermelon (Huang et al., 2016), and pepper (Jiang et al., 2021). Unlike tomato rootstock grafting, which reduced the scion biomass of *C. betacea* seedlings, nightshade rootstock grafting increases the scion biomass, while eggplant rootstock grafting has no significant effect on the scion biomass in cadmium-contaminated soil (Liu et al., 2019). In this experiment, TG and SG had no significant effects on the rootstock root biomass of *C. betacea* seedlings in Se-contaminated soil, while EG decreased the rootstock root biomass. This differed from the results reported in the previous studies (Huang et al., 2016; Liu et al., 2019; Lu et al., 2021; Jiang et al., 2021), indicating that the roots of nightshade, eggplant, and tomato grafted to *C. betacea* scions have different growth responses to cadmium and Se (Liu et al., 2019). Conversely, TG, EG and SG increased the rootstock stem and scion of *C. betacea* seedlings in Se-contaminated soil, which was also inconsistent with the previous studies (Huang et al., 2016; Liu et al., 2019; Jiang et al., 2021; Lu et al., 2021). The scion biomass was indirectly related to the rootstock root biomass, probably due to the growth characteristics of the rootstock (Silva et al., 2018).

Selecting suitable rootstock material for grafting is beneficial in increasing the chlorophyll content of plants (Ren et al., 2019). The grafted seedlings of cucumber and pepper have higher chlorophyll content than their un-grafted counterparts (Fu et al., 2013; Zhang et al., 2019). Grafting also improves the photosynthetic parameters of plants; for example, tomato and apple scion combinations with strong growth potential or high grafting affinity had higher photosynthetic parameters, including the Pn and Gs, which is beneficial for improved photosynthetic efficiency in plants (Zhao et al., 2018; Ren et al., 2019). In this experiment, TG and SG increased chlorophyll a, chlorophyll b, and total chlorophyll content in *C. betacea* seedlings in Se-contaminated soil, and only TG increased the carotenoid content. These results are inconsistent with the previous studies (Zhao et al., 2018; Ren et al., 2019; Liu et al., 2019), probably due to differences in rootstock grow potential (Zhao et al., 2018; Ren et al., 2019). Tomato, eggplant and nightshade are herb plants, and their roots growth faster and have better uptakes of soil nutrients compared with the woody plant *C. betacea* (Wang et al., 2019; Ordoñez-Santos et al., 2020; Lin et al., 2020b). In addition, TG, EG, and SG increased the Pn, Gs, Ci, and Tr of *C. betacea* seedlings in Se-contaminated soil, consistent with the previous studies (Zhao et al., 2018; Ren et al., 2019). Therefore, TG, EG, and SG could improve photosynthesis to promote the growth of *C. betacea* in Se-contaminated soil.

Grafting triggers the accumulation of large amounts of reactive oxygen species (ROS) in plants (Baxter et al., 2014). To regulate the ROS imbalance caused by grafting, rootstocks "activate" the ROS defense system in the scion, and the antioxidant enzymes of the grafted plant act synergistically to improve the antioxidant capacity of the plant (Shahid et al., 2019). Nevertheless, osmoregulatory substances, such as soluble proteins, also regulate and stabilize the structure and function of intracellular biomembranes (Xu et al., 2001; Prajapatia et al., 2023). Studies have reported higher ROS levels or less efficient detoxification systems at incompatible scion/rootstock interfaces (Irisarri et al., 2015). Among the antioxidant systems, SOD plays a major antioxidant defense role, and its enhanced activity induces higher tolerance to oxidative stress (Bowler et al., 1992). In this experiment, TG, EG, and SG increased the SOD and POD activities of *C. betacea* seedlings in Se-contaminated soil, which is consistent with the results of Miao et al. (2019) and Zhang et al. (2010), indicating that grafting could increase the antioxidant enzyme activity of *C. betacea* seedlings in Se-contaminated soil. However, TG, EG, and SG decreased the CAT activity and soluble protein content of *C. betacea* seedlings in Se-contaminated soil. This is inconsistent with the previous studies (Zhang et al., 2010; Miao et al., 2019), probably because the antioxidant capacity of grafted plants is directly related to the rootstock-scion affinity and higher ROS accumulation after grafting inhibits CAT activity (Bai, 2019).

There are many measures that can affect the element absorption of plants (Durga et al., 2017; Agnihotri et al., 2018; Kim et al., 2024; Patel et al., 2024; Kumar et al., 2024; Mariyam et al., 2024). Among these measures, grafting changes plant roots and can alter the absorption and transport of essential ions in plants, thereby affecting the nutritional level and accumulation of other elements in the aboveground parts of plants (Duan et al., 2017). Plant absorption and transport of Se are mainly determined by the form of Se insoil and the genetic background of plants (Sors et al., 2005), and plant roots absorb selenite through the phosphate-transport protein (Li et al., 2008). In this experiment, the Se content in the rootstock root, rootstock stem, and scion of *C. betacea* seedlings exhibites the rootstock root > rootstock stem > scion increment pattern. This is because tetravalent Se (Ⅳ) is converted into hexavalent Se (Ⅵ) and organic Se compounds before being transported to the aboveground parts, and the absorption rate of Se is greater than the conversion rate (Zayed et al., 1998). In this experiment, TG, EG, and SG decreased the Se content of the rootstock roots and rootstock stems of *C. betacea* seedlings. This was inconsistent with the results of Hu et al. (2021), probably due to the different Se accumulation capacities of the rootstocks (Wang et al., 2019). SG increased the scion Se content and BCF of *C. betacea* seedlings, while TG and EG decreased the Se content and BCF. Moreover, TG, EG, and SG increased the $TF_{shoot/rootstock\ root}$ and $TF_{rootstock\ stem/rootstock\ root}$ of *C. betacea* seedlings, while only SG increased the $TF_{scion/rootstock\ stem}$ of *C. betacea* seedlings. These results indicate that the grafting promotes the Se transport from the roots to shoots while the TG and EG inhibites and SG promotes Se transport from the rootstock stems to scions. This phenomenon is related to the grafting-enhanced stress resistance in plants (Lin et al., 2020b; Zhang, 2021). Therefore, the interaction between the rootstock and scion in SG enhances the Se uptake in *C. betacea* seedlings. Moreover, PCA and cluster analysis showed that the root biomass, scion Se content, CAT activity, soluble protein content, rootstock root Se content, and rootstock stem Se content cluster into one category, indicating these parameters were closely related to the scion Se content. However, the interaction mechanisms between the three rootstocks and *C. betacea* scions need further studies.

5 Conclusion

Tomato, eggplant, and nightshade rootstock graftings promote the growth of *C. betacea* seedlings in Se-contaminated soil and improve the photosynthesis and antioxidant enzyme activities of *C. betacea* seedlings. Nightshade rootstock grafting also promotes Se uptake in *C. betacea* seedlings, while tomato and eggplant rootstocks graftings decrease it. In addition, root biomass, catalase activity, soluble protein content, rootstock root Se content, and rootstock stem Se content play important roles on the scion Se content. Therefore, nightshade can be considered as the suitable rootstock for enhancing the growth and Se uptake by *C. betacea*. Future studies should explore the grafting effects and Se uptake by *C. betacea* fruits grafted on nightshade rootstock and how nightshade rootstock promotes Se uptake by *C. betacea* scions.

References

[1] BAI X. Study on the affinity physiological mechanisms during rootstock healing of thin-skinned melon grafting[D]. Guangxi University, Nanning, China, 2019.

[2] BAXTER A, MITTLER R, SUZUKI N. ROS as key players in plant stress signaling [J]. Journal of Experimental Botany, 2014, 65(5):1229-1240.

[3] BOWLER C, MONTAGU M V, Inze D. Superoxide dismutase and stress tolerance [J]. Annual Review of Plant Physiology and Plant Molecular Biology, 1992, 43(1):83-116.

[4] CHEN S P, ZHOU G S, LIU X G, et al. Effects of different rootstocks on the yield, the content of xylem exudation and minerals of tomato[J]. Journal of Shanxi Agricultural University (Natural Science Edition), 2019, 39(5):62-67.

[5] DUAN Y, XIANG C G, HU J. Effects of pumpkin rootstock on silicon transport and

distribution in shoots of grafted cucumber[J]. China Cucurbits and Vegetables, 2017, 30(10):6-11.

[6] FU L, BAI X M, YANG X H, et al. Effects of grafted on photosynthetic characteristics, yield and quality of pepper [J]. Acta Horticulturae Sinica, 2013, 40(3):449-457.

[7] GUO S Y. Study on the cultivation effect of grafting tomatoes on different rootstocks[D]. Guangxi University, Nanning, China, 2018.

[8]GUPTA M, GUPTA S. An overview of selenium uptake, metabolism, and toxicity in plants[J]. Frontiers in Plant Science, 2017, 7(1):2074.

[9]HAO Z B, CANG J, XU Z. Plant physiology experiment[M]. Harbin Institute of Technology Press, Harbin, China, 2004.

[10]HE W, PAN H L, PAN T F, et al. Research progress on the interaction between scion and rootstock in fruit trees[J]. Acta Horticulturae Sinica, 2017, 44(9):1645-1657.

[11] HU R P, FAN Z H, CHEN Q H. Effects of *Solanum nigrum* rootstock with different ploidies grafting on selenium accumulation of *Cyphomandra betacea* seedlings[J]. Chinese Agricultural Science Bulletin, 2021, 37(2):43-47.

[12]HUANG K W, LI K Q, SUI L Y, et al. Effects of different number of grafting on physiological characteristics and selenium absorption of *Nasturtium officinale* cuttings[J]. Acta Agriculturae Borcali-Sinica, 2020, 29(3):419-427.

[13] HUANG Y, JIAO Y Y, NAWAZ M A, et al. Improving magnesium uptake, photosynthesis and antioxidant enzyme activities of watermelon by grafting onto pumpkin rootstock under low magnesium[J]. Plant and Soil, 2016, 409(1-2):229-246.

[14] IRISARRI P, BINCZYCKI P, ERREA P, et al. Oxidative stress associated with rootstock-scion interactions in pear/quince combinations during early stages of graft development[J]. Journal of Plant Physiology, 2015, 176:25-35.

[15]JIANG X M, TIAN X, CHEN Y T. Effects of both-root-cut grafting on the wound healing and growth of grafting seedling of pepper [J]. Northern Horticulture, 2021, (2):10-16.

[16]JOZWIAK W, POLITYCKA B. Effect of selenium on alleviating oxidative stress caused by a water deficit in cucumber roots[J]. Plants, 2019, 8(7):217.

[17]KHALID M F, MORILLON R, ANJUM M A, et al. Volkamer lemon tetraploid rootstock transmits the salt tolerance when grafted with diploid Kinnow mandarin by strong

antioxidant defense mechanism and efficient osmotic adjustment[J]. Journal of Plant Growth Regulation, 2022, 41(3):1125-1137.

[18] KHANKAHDANI H H, RASTEGAR S, GOLEIN B, et al. Effect of rootstock on vegetative growth and mineral elements in scion of different Persian lime (*Citrus latifolia* Tanaka) genotypes[J]. Scientia Horticulturae, 2019, 246:136-145.

[19] KIM D Y, YANG T, SRIVASTAVA P, et al. Alginic acid-functionalized silver nanoparticles: A rapid monitoring tool for detecting the technology-critical element tellurium [J]. Journal of Hazardous Materials, 2024, 465:133161.

[20] KUMAR D, SINGH R, UPADHYAY S K, et al. Review on interactions between nanomaterials and phytohormones: Novel perspectives and opportunities for mitigating environmental challenges[J]. Plant Science, 2024, 340:111964.

[21] LANZA M G D B, REIS A R D. Roles of selenium in mineral plant nutrition: ROS scavenging responses against abiotic stresses[J]. Plant Physiology and Biochemistry, 2021, 164:27-43.

[22] LI H F, MCGRATH S P, ZHAO F J. Selenium uptake, translocation and speciation in wheat supplied with selenate or selenite[J]. The New Phytologist, 2008, 178(1):92-102.

[23] LI Z Y, FAN R, PENG X M. Salicylic acid alleviates selenium stress and promotes selenium uptake of grapevine[J]. Physiology and Molecular Biology of Plants, 2022, 28(3): 625-635.

[24] LIN L J, LI Z Y, WANG J, et al. 24-epibrassinolide promotes selenium uptake in grapevine under selenium stress[J]. Scientia Horticulturae, 2023, 308:111564.

[25] LIN L J, SUN J, CUI T H, et al. Selenium accumulation characteristics of *Cyphomandra betacea* (*Solanum betaceum*) seedlings[J]. Physiology and Molecular Biology of Plants, 2020a, 26(7):1375-1383.

[26] LIN L J, WANG J Y, LIAO M A, et al. *Artemisia argyi* water extract promotes selenium uptake of peach seedlings[J]. Frontiers in Plant Science, 2022, 13:1014454.

[27] LIN L J, WU C F, JIANG W, et al. Grafting increases cadmium accumulation in the post-grafting generations of the potential cadmium-hyperaccumulator *Solanum photeinocarpum*[J]. Chemistry and Ecology, 2020b, 36(7):685-704.

[28] LIN L J, YANG D Y, WANG X, et al. Effects of grafting on the cadmium accumulation characteristics of the potential Cd-hyperaccumulator *Solanum photeinocarpum* [J]. Environmental Monitoring and Assessment, 2016, 188(2):82.

[29] LIU Q, HUO R, LIN L J, et al. Effects of different rootstocks on cadmium accumulation of grafted *Cyphomandra betacea* seedlings [J]. International Journal of Environmental Analytical Chemistry, 2019, 99(12):1247-1254.

[30] LU R H, CUI Y L, CHENG Z Q, et al. Effects of different rootstocks grafting on selenium accumulation in eggplant[J]. Northern Horticulture, 2021, (6):1-7.

[31] LUO H Z. Effects of mutual grafting on the growth and selenium enrichment of cherry tomato progeny of two genotypes [D]. Sichuan Agricultural University, Chengdu, China, 2019.

[32] MAURYA D, PANDEY A K, KUMAR V, et al. Grafting techniques in vegetable crops: A review[J]. International Journal of Chemical Studies, 2019, 7(2):1664-1672.

[33] MIAO L, LI S Z, BAI L Q, et al. Effect of grafting methods on physiological change of graft union formation in cucumber grafted onto bottle gourd rootstock[J]. Scientia Horticulturae, 2019, 244:249-256.

[34] NOGUEIRA M L, CARVALHO M E A, FERREIRA J M M, et al. Cadmium-induced transgenerational effects on tomato plants: A gift from parents to progenies [J]. Science of The Total Environment, 2021, 789:147885.

[35] ORDOñEZ-SANTOS L E, MARTINEZ-GIRON J. Thermal degradation kinetics of carotenoids, vitamin C and provitamin A in tree tomato juice [J]. International Journal of Food Science and Technology, 2020, 55(1):201-210.

[36] PRAJAPATIA P, GUPTA P, KHARWAR R N, et al. Nitric oxide mediated regulation of ascorbate-glutathione pathway alleviates mitotic aberrations and DNA damage in *Allium cepa* L. under salinity stress [J]. International Journal of Phytoremediation, 2023, 25 (4):403-414.

[37] RASTMANESH F, MOORE F, KESHAVARZI B. Speciation and phytoavailability of heavy metals in contaminated soils in Sarcheshmeh area, Kerman Province, Iran [J]. Bulletin of Environmental Contamination and Toxicology, 2010, 85(5):515-519.

[38] REN P, CHANG Q, FU B, et al. Effects of rootstock-scion interaction on tomato photosynthesis, growth and yield[J]. Plant Physiology Journal, 2019, 55(9):1325-1334.

[39] ROMAN M, JITARU P, BARBANTE C. Selenium biochemistry and its role for human-health[J]. Metallomics: Integrated Biometal Science, 2014, 6(1):25-54.

[40] SHAHID M A, BALAL R M, KHAN N, et al. Rootstocks influence the salt tolerance of Kinnow mandarin trees by altering the antioxidant defense system, osmolyte

concentration, and toxic ion accumulation[J]. Scientia Horticulturae, 2019, 250:1-11.

[41]SILVA S L, DINIZ R P, NEVES R D J, et al. Grafting as a strategy to increase flowering of cassava[J]. Scientia Horticulturae, 2018, 240:544-551.

[42]SORS T G, ELLIS D R, SALT D E. Selenium uptake, translocation, assimilation and metabolic fate in plants[J]. Photosynthesis research, 2005, 86(3):373-389.

[43] WANG X, HUANG Z, LIN L J, et al. Study on selenium accumulation characteristics of *Lycopersicon esculentum*, *Solanum melongena* and *Solanum nigrum*[J]. IOP Conference Series:Earth and Environmental Science, 2019, 310(4):042065.

[44] XU Q S, SHI G X, DU K H. Effect of Cd^{2+} on antioxidese system and ultrastructure of *Ottelia alismoides* (L.)Pers. leaves[J]. Guihaia, 2001, 22(1):92-96.

[45]ZAYED A, LYTLE C M, TERRY N. Accumulation and volatilization of different chemical species of selenium by plants[J]. Planta, 1998, 206(2):284-292.

[46]ZHANG H M, JIN H J, DING X T, et al. Physiological response and resistance evaluation of grafted cucumber seedlings with different rootstocks under temperature stress [J]. Acta Botanica Boreali-Occidentalia Sinica, 2019, 39(7):1259-1269.

[47]ZHANG P, XUE J L, LIU Y, et al. Research progress on responses and tolerance mechanisms of plants to selenium pollution[J]. Journal of Anhui Agricultural Sciences, 2023, 51(17):14-17.

[48] ZHANG Z H. Study on the mechanism of different rootstocks and scions combinations regulating drought tolerance of grafted tomato seedling[D]. Shandong Agricultural University, Taian, China, 2021.

[49]ZHANG Z K, LIU S Q, HAO S Q, et al. Grafting increases the copper tolerance of cucumber seedlings by improvement of polyamine content and enhancement of antioxidant enzymes activity[J]. Agricultural Sciences in China, 2010, 9(7):985-994.

[50]ZHAO T, CHENG L, WANG C, et al. Effect of different apples cion-rootstock combinations on growth and photosynthesis characteristics[J]. Acta Botanica Boreali-Occidentalia Sinica, 2018, 38(9):1707-1716.

Section Ⅱ Different ploidies of *Solanum nigrum* rootstocks

1 Introduction

Selenium (Se) is a non-heavy metal element and one of the essential trace elements necessary for the growth and development of humans and animals (Li et al., 2019). China is a typical Se-deficient country, with 50% of the population having insufficient Se intake to varying degrees (Sun et al., 2017). Increasing Se intake appropriately can have a significant impact on enhancing national resistance and effectively preventing diseases caused by Se deficiency (Bai et al., 2020). Currently, there are studies on increasing Se content in crops through Se supplementation from soil. However, it has been noted that most agricultural products are non-Se-enriched plants, meaning they are poor in Se enrichment and transport (Zhou et al., 2014; Thiry et al., 2012). Therefore, it is of practical significance to improve the ability of crops to enrich Se through appropriate cultivation techniques or agronomic measures.

Grafting is an asexual propagation technique in which the rootstock is connected to the scion and develops into a new plant through the formation of healing tissue. It is widely used in cultivation due to its advantages such as restoring tree species and promoting growth to breed virus-free varieties (Luo et al., 1996; Zeist et al., 2020). In addition to maintaining the excellent characteristics of the parent plant, grafting also enhances the plant's resistance to insects (Liang, 2012), diseases (Liu et al., 2013), drought (Zhou et al., 2015), and coldness (Huang et al., 2013). Recent studies have shown that grafting with different rootstocks can affect the scion's uptake of mineral elements (Liu et al., 2012; Zhang et al., 2014). Zhang et al. (2014) found that grafting with dwarfing rootstocks could promote the absorption of phosphorus, potassium, calcium, and copper elements in apples; Ji et al. (2014) found that grafting apples with Fuping tulip tree as a rootstock under low-phosphorus conditions was conducive to enhancing phosphorus uptake and utilization efficiency in apples; and Huang et al. (2020b) found that grafting *Nasturtium officinale* on rootstock broccoli could promote the accumulation of Se in *N. officinale*. Therefore, grafting on different rootstocks can alter the nutrient absorption of plants and, to a certain extent, can also change the absorption and accumulation of Se.

Solanum nigrum and *Solanum photeinocarpum* are both annual herbaceous plants in the Solanaceae family, known for their high stress resistance and ability to enrich and transport Se (Huang et al., 2018; Wu et al., 2018). *S. nigrum* is hexaploid while *S. photeinocarpum* is diploid, with *S. nigrum* believed to have evolved from *S. photeinocarpum* (Zhang et al., 2019; Xu et al., 2004; Tandon et al., 1964). *Cyphomandra betacea*, a perennial evergreen shrub in the Solanaceae family, produces nutritious, sweet, sour, and juicy fruits with high ornamental and edible value (Zhi et al., 2014). In this study, *S. photeinocarpum* (diploid), tetraploid progeny of *S. photeinocarpum* induced by colchicine, and *S. nigrum* (hexaploid) were used as rootstocks, with *C. betacea* as the scion to investigate the impact of grafting *S. nigrum* with different ploidies on Se accumulation in *C. betacea*. The goal of this research was to identify the most effective rootstock for promoting the growth and Se enrichment of *C. betacea*, ultimately providing insights for producing Se-enriched *C. betacea*.

2 Materials andmethods

2.1 Materials

C. betacea seeds were collected from a five-year-old fruiting tree at the Chengdu Campus of Sichuan Agricultural University (30°42′N, 103°51′E) in September 2019. The seeds were air-dried and stored at 4 ℃. The seeds of *S. photeinocarpum* (diploid) and *S. nigrum* (hexaploid) were obtained from the farmlands around the Chengdu Campus. The shoot tips of the diploid seedlings at the stage of unfolded cotyledons were treated with colchicine (0.3%) for 48 h, and the tetraploid was induced. Tetraploid seeds were obtained from the diploid plants treated with colchicine. Diploid, tetraploid, and hexaploid seeds were air-dried and stored at 4 ℃.

The soil used in this experiment was collected from thefarmland around the Chengdu Campus. The soil was fluvo-aqic, and the basic phyicochemical properties of soil samples are presented in Table 4-4.

Table 4-4 The basic chemical properties of soil

Soil type	pH value	Organic matter/ (g/kg)	Total N/ (g/kg)	Total P/ (g/kg)	Total K/ (g/kg)	Alkaline hydrolyzed N/ (mg/kg)	Available P/ (mg/kg)	Available K/ (mg/kg)	Total Se/ (mg/kg)
Fluvo-aquic	7.58	28.02	1.44	0.43	24.55	58.68	19.23	58.33	0.35

2.2 Experimental design

In September 2019, the seeds of four plants were planted in seedling trays filled with wet perlite with the following conditions: 14-h days at 25 ℃, relative humidity 75%, 4,000 Lux; and 10-h nights at 20 ℃, relative humidity 95%, 0 Lux. Hoagland nutrient solutions were used to irrigate the seedlings every three days after emergence. When plant seedlings reached a height of 10 cm, grafting occurred. The scions of *C. betacea* seedlings were grafted onto the rootstocks of different ploidies of *S. nigrum* seedlings. The grafting method and post-grafting management of plants were conducted as described by Lin et al. (2016).

The soil samples were treated in September 2019, air-dried, and passed through 6.72-mm mesh as described by Lin et al. (2020). Each plastic pot (15 cm height, 18 cm diameter) contained 3.0 kg of air-dried soil samples. Na_2SeO_3 was added to the soil to make the final soil Se concentration of 5 mg/kg. The soil moisture content was maintained at 80% of field capacity for one month. In October 2019 (15 days after grafting), the surviving grafted seedlings were transplanted into pots containing Se-contaminated soil with each pot containing three plants. The treatments included un-grafted (CK), *S. photeinocarpum* grafting (diploid), tetraploid *S. photeinocarpum* grafting (tetraploid), and *S. nigrum* grafting (hexaploid). Each treatment was conducted in triplicate, and the soil moisture content was maintained at 80% of the field capacity. After 40 days of growth in Se-contaminated soils, the plants were harvested.

2.3 Determination of indicators

The fifth mature leaves from the top of the plants were also used to measure the content of photosynthetic pigments (chlorophyll a, chlorophyll b, total chlorophyll and carotenoid). Photosynthetic pigment was extracted using the acetone-ethanol extraction method and determined at 663, 645, 652, and 470 nm wavelengths using a spectrophotometer (Summit, Shanghai, China) according to reports of Xiong (2003). The Chlorophyll a/b was calculated as chlorophyll a/b content. Then, the plants were harvested, and divided into rootstock root, rootstock stem, scion stem, and scion leaf. Each part was washed with tap water followed by deionized water, and then dried at 80 ℃ until constant weight was achieved to determine their dry biomass. For the CK plants, the stem was collected, and the lower part of the stem (with the same height as the grafted plants) was treated as the rootstock, while the upper stem with the leaves was treated as the scion. Thus, the rootstock root, rootstock stem, scion stem, and scion leaf biomass were determined using an electronic balance, and the whole scion biomass was calculated as scion stem biomass + scion leaf biomass. The dried samples were ground and sieved through a 0.15-mm mesh nylon sieve. Then, the ground samples were digested with 4:1 (v/v) nitrate-perchloric acid and reduced with hydrochloric acid. The solutions were used to determine the Se content. The Se content in rootstock roots, rootstock stems and scions was determined using a hydride generation-atomic fluorescence spectrometry (AFS-9700, Beijing Haiguang Instrument Co., Ltd., Beijing, China). The whole scion Se content was calculated as (scion stem Se content × scion stem biomass + scion leaf Se content × scion leaf biomass)/ whole scion biomass, Se accumulation amount was calculated as Se content × biomass, and the whole scion Se accumulation amount was calculated as scion stem Se accumulation amount + scion leaf Se accumulation amount (Li et al., 2008).

2.4 Statistical analyses

Statistical analyses were conducted using SPSS 20.0 software (IBM, Chicago, IL, USA). Data was normalized and subjected to a homogeneity test before a one-way analysis of variance and Duncan's multiple range test ($P < 0.05$).

3 Results

3.1 Biomass

Compared with CK, the treatments of tetraploid and hexaploid increased the rootstock stem, scion stem, scion leaf, and whole scion biomass of *C. betacea* seedlings, while had no significant effect on the root biomass (Table 4-5). The treatment of hexaploid had the highest biomass, which increased the rootstock stem, scion stem, scion leaf, and whole scion biomass by 86.27%, 57.39%, 71.83%, and 69.54%, respectively, compared with CK. The treatment of diploid decreased the rootstock root biomass, increased the rootstock stem biomass, and did not affect the scion stem, scion leaf, and whole scion biomass, compared with CK.

Table 4-5　Biomass of *C. betacea* seedlings

Treatment	Rootstock root/ (g/plant)	Rootstock stem/ (g/plant)	Scion stem/ (g/plant)	Scion leaf/ (g/plant)	Whole scion/ (g/plant)
CK	0. 200±0. 005a	0. 102±0. 003d	0. 115±0. 006c	0. 536±0. 014c	0. 650±0. 020c
Diploid	0. 134±0. 006b	0. 118±0. 002c	0. 119±0. 002bc	0. 540±0. 010c	0. 659±0. 011c
Tetraploid	0. 193±0. 008a	0. 137±0. 005b	0. 128±0. 004b	0. 672±0. 011b	0. 801±0. 015b
Hexaploid	0. 204±0. 009a	0. 190±0. 006a	0. 181±0. 003a	0. 921±0. 016a	1. 102±0. 018a

Note: Different lowercase letters indicate significant differences among the treatments (Duncan's multiple range test, $P < 0.05$).

3. 2　Photosynthetic pigment content

The treatments of diploid, tetraploid, and hexaploid increased the content of chlorophyll a, chlorophyll b, and total chlorophyll in *C. betacea* seedlings (Table 4-6). The treatment of hexaploid had the maximum chlorophyll content, and increased the content of chlorophyll a, chlorophyll b, and total chlorophyll by 49.76%, 51.20%, and 50.11%, respectively, compared with CK. The treatments of diploid, tetraploid, and hexaploidy did not affect the ratio of chlorophyll a/b. For the carotenoid content, the treatments of tetraploid and hexaploid increased it, while the treatment of diploid did not affect it.

Table 4-6　Photosynthetic pigment content in *C. betacea* seedlings

Treatment	Chlorophyll a/ (mg/g)	Chlorophyll b/ (mg/g)	Total chlorophyll/ (mg/g)	Chlorophyll a/b	Carotenoid/ (mg/g)
CK	1. 035±0. 030c	0. 332±0. 017c	1. 367±0. 046c	3. 119±0. 066a	0. 221±0. 003c
Diploid	1. 350±0. 037b	0. 414±0. 013b	1. 764±0. 050b	3. 261±0. 011a	0. 232±0. 006c
Tetraploid	1. 492±0. 013a	0. 467±0. 015a	1. 960±0. 029a	3. 195±0. 077a	0. 293±0. 005b
Hexaploid	1. 550±0. 023a	0. 502±0. 020a	2. 052±0. 044a	3. 092±0. 078a	0. 328±0. 002a

Note: Different lowercase letters indicate significant differences among the treatments (Duncan's multiple range test, $P < 0.05$).

3. 3　Se content

The treatments of diploid, tetraploid, and hexaploid increased the Se content in rootstock roots, scion stems, scion leaves, and whole scions of *C. betacea* seedlings, and the order was ranked as hexaploid > tetraploid > diploid > CK (Table 4-7). For the Se content in rootstock stems, the treatments of tetraploid and hexaploid increased it, and increased by 69. 56% and

120.57%, respectively, compared with CK. The treatment of diploid did not affect the Se content in rootstock stems.

Table 4-7 Se content in *C. betacea* seedlings

Treatment	Rootstock root/(mg/kg)	Rootstock stem/(mg/kg)	Scion stem/(mg/kg)	Scion leaf/(mg/kg)	Whole scion/(mg/kg)
CK	11.22±0.08d	2.625±0.116c	2.197±0.087d	2.483±0.020d	2.433±0.031d
Diploid	12.32±0.40c	2.717±0.073c	2.722±0.091c	2.874±0.050c	2.847±0.024c
Tetraploid	13.99±0.22b	4.451±0.128b	3.668±0.085b	3.159±0.054b	3.241±0.031b
Hexaploid	18.15±0.43a	5.790±0.162a	4.251±0.077a	3.719±0.092a	3.806±0.090a

Note: Different lowercase letters indicate significant differences among the treatments (Duncan's multiple range test, $P < 0.05$).

3.4 Se accumulation amount

The treatment of diploid decreased the Se accumulation amount in rootstock roots of *C. betacea* seedlings, and did not affect the Se accumulation amount in rootstock stems (Table 4-8). The treatments of tetraploid and hexaploid increased the Se accumulation amounts in rootstock roots and rootstock stems. The treatments of diploid, tetraploid, and hexaploid increased the Se accumulation amounts in scion stems, scion leaves, and whole scions. The treatment of hexaploid had the best effect on the increasing of Se accumulation amount. Compared with CK, the treatment of hexaploid increased the Se accumulation amounts in scion stems, scion leaves, and whole scions by 205.56%, 157.52%, and 165.17%.

Table 4-8 Se accumulation amount in *C. betacea* seedlings

Treatment	Rootstock root/(μg/plant)	Rootstock stem/(μg/plant)	Scion stem/(μg/plant)	Scion leaf/(μg/plant)	Whole scion/(μg/plant)
CK	2.248±0.036c	0.268±0.019c	0.252±0.023d	1.330±0.045d	1.582±0.068d
Diploid	1.647±0.133d	0.320±0.013c	0.323±0.015c	1.553±0.001c	1.877±0.016c
Tetraploid	2.700±0.150b	0.612±0.041b	0.470±0.024b	2.125±0.002b	2.595±0.022b
Hexaploid	3.705±0.073a	1.101±0.066a	0.770±0.026a	3.425±0.143a	4.195±0.169a

Note: Different lowercase letters indicate significant differences among the treatments (Duncan's multiple range test, $P < 0.05$).

4 Discussion

Grafting is a traditional agronomic measure that holds significant value in increasing plant yield, enhancing stress resistance, and altering material absorption and accumulation traits (Lu et al., 2008; Lin et al., 2015). The efficacy of grafting on plant growth is closely linked to rootstock selection, with the choice of appropriate rootstocks proving pivotal in boosting the biomass of grafted seedlings (Tan et al., 2014; Yuan et al., 2019). The results of the experiment indicated that grafting treatments using *S. nigrum* of varying ploidies as rootstocks had diverse impacts on the growth of *C. betacea* seedlings. The effect varied depending on the ploidy of *S. nigrum* rootstocks. Employing tetraploid and hexaploid as rootstocks for grafting notably increased the biomass of the rootstock stem, scion stem, scion leaf, and overall scion of *C. betacea* seedlings. However, it had no significant influence on root biomass, suggesting that using tetraploid and hexaploid as rootstocks for grafting favored the above-ground growth of *C. betacea*. This could be attributed to the thicker stems, leaves, and sturdier plants of polyploid rootstocks, which stimulate the growth of grafted seedlings (Dang et al., 2019; Allario et al., 2011).

Photosynthetic pigments are essential for photosynthesis in plants. The selection of suitable rootstocks can help increase the content of photosynthetic pigments in grafted seedlings. These pigments are converted into chemical energy by plants through photosynthesis, which is then used for plant growth and development (Tan et al., 2014). In this experiment, grafting with both tetraploid and hexaploid rootstocks significantly increased the content of chlorophyll a, chlorophyll b, and carotenoid in *C. betacea* seedlings. The biomass and photosynthetic pigment content of *C. betacea* seedlings reached the maximum when the rootstock was hexaploid, indicating that the effect of grafting on *C. betacea* may be related to photosynthetic pigment content. These results are in line with previous grafting experiments in cucumber (Zhang et al., 2019), watermelon (Yang et al., 2015), and pepper (Fu et al., 2013).

The results of this experiment indicate that the grafting of different ploidies of *S. nigrum* as rootstocks haS varying degrees of influence on Se content and Se accumulation in *C. betacea* seedlings. Grafting with both tetraploid and hexaploid rootstocks increased the Se content in roots, stems, leaves, and whole scions of *C. betacea*. This suggests that interactions between tetraploid and hexaploid rootstocks promote Se uptake and accumulation in *C. betacea*. This may be attributed to the influence of grafting on the expression of Se transporter-related proteins and metabolic pathways in *C. betacea* (Huang et al., 2020b). Se accumulation is indicative of a plant's ability to extract Se from the soil, which is crucial for plant growth and Se uptake capacity (Huang et al., 2020a; Wu et al., 2018). Grafting with diploid rootstocks decreased Se accumulation in *C. betacea* rootstock roots, while all other grafting treatments increased Se accumulation in rootstock roots. Grafting with different ploidies of *S. nigrum* rootstocks also increased Se accumulation in stems, leaves, and whole scions of *C. betacea* seedlings. Among the different rootstocks, the hexaploid rootstock had the most significant effect, likely due to the Se-enriched nature of *S. nigrum* enhancing the growth and Se transport capacity of *C. betacea* scion (Huang et al., 2018). These findings align with similar results observed in *N. officinale* from cuttings grafted with different rootstocks by Huang et al. (2020a).

5 Conclusion

The different rootstock combinations affect the growth and Se accumulation of *C. betacea*. Grafting rootstocks with hexaploid as rootstocks has the best effect on increasing biomass, photosynthetic pigment content, Se content, and Se accumulation of *C. betacea*. This could provide a certain reference for Se-enriched *C. betacea* cultivation. However, the mechanism of Se enrichment in *C. betacea* through grafting has not been addressed in this study and needs to be further explored in the future.

References

[1] ALLARIO T, JAVIER B, COLMENERO-FLORES J M, et al. Large changes in anatomy and physiology between diploid Rangpur lime (*Citrus limonia*) and its autotetraploid are not associated with large changes in leaf gene expression[J]. Journal of Experimental Botany, 2011, 62(8):2507-2519.

[2] BAI X, SUN M, DONG S F, et al. Effects of selenium supplementation in Se-enriched areas on selenium accumulation of triticale plants[J]. Journal of Hebei Agricultural University, 2020, 43(2):34-39.

[3] DANG J B, LIANG G L, LI C, et al. Polyploid rootstock of fruit tree: research status and prospects[J]. Acta Horticulturae Sinica, 2019, 46(9):1701-1710.

[4] FU L, BAI X M, YANG X H, et al. Effects of grafted on photosynthetic characteristics, yield and quality of pepper [J]. Acta Horticulturae Sinica, 2013, 40

(3):449-457.

[5]HUANG K W, LI K Q, LIU J, et al. Effects of different rootstock combinations on physiological characteristics and selenium accumulation of *Nasturtium officinale* cuttings[J]. Acta Agriculturae Zhejiangensis, 2020a, 32(3):447-454.

[6]HUANG K W, LI K Q, SUI L Y, et al. Effects of different number of grafting on physiological characteristics and selenium absorption of *Nasturtium officinale* cuttings[J]. Acta Agriculture Boreali-occidentalis Sinica, 2020b, 29(3):419-427.

[7]HUANG K, LIN L, LIAO M. Study on selenium accumulation characteristics of *Solanum nigrum*[J]. IOP Conference Series: Earth and Environmental Science, 2018, 199(4):42027.

[8]HUANG Q C, ZHAO H T, LIU J M, et al. Effects of grafting with seedling on cytoplasmic male sterile lines and its maintainers in kenaf[J]. Journal of Huazhong Agricultural University, 2013, 32(3):8-14.

[9]JI M M, XU H G, PENG L, et al. Characteristics of phosphorus absorption and utilization in five apple rootstocks under low phosphorus stress[J]. Journal of Plant Nutrition and Fertilizer, 2014, 20(4):974-980.

[10]LI L P, FAN W, HUANG Z Y, et al. Total selenium and se species in vegetables measured with hydride generation-atomic fluorescence spectrometry[J]. Spectroscopy and Spectral Analysis, 2008, 28(12):2975-2978.

[11]LI L Q, DENG P, HUANG C D, et al. Research on selenium enrichment ability of different waxy corn varieties[J]. Journal of Anhui Agricultural Sciences, 2019, 47(21):157-159.

[12]LIANG P. Study on the resistance evaluation of grafted tomato (*Lycopersicon esculentum* Mill.) seedlings and resistance mechanisms to Meloidogyne incognita[D]. Southwest University, Chongqing, China, 2012.

[13]LIN L J, LUO L, ZHANG X, et al. Sichuan agricultural effects of rape rootstock on cadmium accumulation characteristics of *Capsella bursa-pastoris* post-grafting generation[J]. Acta Agriculture Boreall-Sinica, 2015, 30(1):207-212.

[14]LIN L J, WU C F, JIANG W, et al. Grafting increases cadmium accumulation in the post-grafting generations of the potential cadmium-hyperaccumulator *Solanum photeinocarpum*[J]. Chemistry and Ecology, 2020, 36(7):685-704.

[15]LIN L J, YANG D Y, WANG X, et al. Effects of grafting on the cadmium

accumulation characteristics of the potential Cd-hyperaccumulator *Solanum photeinocarpum* [J]. Environmental Monitoring and Assessment, 2016, 188(2):82.

[16] LIU Q, WEI M, SHEN Q, et al. Effects of different rootstocks on bloom formation and absorption and distribution of silicon in grafted cucumber[J]. Acta Horticulturae Sinica, 2012, 39(5):897-904.

[17] LIU Y X, FU L, AI X Z, et al. Effect of grafting on secondary metabolism and its relationship with bacterial wilt resistance in pepper[J]. Scientia Agricultura Sinica, 2013, 46(14):2963-2969.

[18] LU M M, SUN S. Effect of cadmium stress on growth and membrane-lipid peroxidation of small water melon seedlings[J]. Journal of Shanxi Agricultural Sciences, 2008, 36(12):64-66.

[19] Luo Z R, Huo G, Cai L H. The roles of grafting in plant propagation and improvement[J]. Plant Physiology and Agriculture, 1996(1):59-63.

[20] SUN G X, LI Y, LI G, et al. Climatic causes of the selenium-deficient soil belt inChina[J]. Current Biotechnology, 2017, 7(5):387-394.

[21] TAN M M, ZHANG X Y, Fu Q S, et al. Effects of grafting on physiological characteristics of melon (*Cucumis melo*)seedlings under copper stress[J]. Chinese Journal of Applied Ecology, 2014, 25(12):3563-3572.

[22] TANDON S L, RAO G R. Cytogenetical investigations in relation to the mechanism of evolution in hexaploid*Solanum nigrum* L. [J]. Nature, 1964, 201(4926):1348-1349.

[23] THIRY C, RUTTENS A, TEMMERMAN L D, et al. Current knowledge in species-related bioavailability of selenium in food [J]. Food Chemistry, 2012, 130(4):767-784.

[24] WU C F, BAO R F, LIAO H P, et al. Selenium accumulation characteristics of wild vegetable *Solanum photeinocarpum*[J]. Journal of Sichuan Agricultural University, 2018, 36(6):778-784.

[25] XIONG Q E. Plant physiology experiment tutorial[M]. Sichuan Publishing House of Science and Technology, Chengdu, 2003.

[26] XU X F, ZHANG H Y, YUAN Q H, et al. Studies on chromosome of three types of *Solanum nigrum* in China[J]. Guihaia, 2004, 24(6):544-545.

[27] YANG Y, YU L, WANG L, et al. Bottle gourd rootstock-grafting promotes

photosynthesis by regulating the stomata and non-stomata performances in leaves of watermelon seedlings under NaCl stress [J]. Journal of Plant Physiology, 2015, 168-187:50-58.

[28] YUAN J W, LI M M, YIN Y G, et al. Physiological responses of *Cabernet sauvian* seedling grafted on different salt-tolerant rootstocks to NaCl stress[J]. Journal of Agriculture University, 2019, 24(8):48-59.

[29] ZEIST A R, DE RESENDE J T V, ZANIN D S, et al. Effect of acclimation environments, grafting methods and rootstock RVTC-66 on the seedling development and production of tomato[J]. Scientia Horticulturae, 2020, 271:109496.

[30] ZHANG H M, JIN H J, DING X T, et al. Physiological response and resistance evaluation of grafted cucumber seedlings with different rootstocks under temperature stress [J]. Acta Botanica Boreali-Occidentalia Sinica, 2019, 39(7):1259-1269.

[31] ZHANG X Z, GUO J Y, WANG Y Z, et al. Effects of different rootstocks on mineral content and fruit qualities of Fuji apple[J]. Journal of Plant Nutrition and Fertilizer, 2014, 20(2):414-420.

[32] ZHI S, WU Y X, WANG Y J, et al. Establishing high-frequency regeneration system of *Cyphomandra betacea*[J]. Hubei Agricultural Sciences, 2014, 53(23):5880-5883.

[33] ZHOU Q K, ZHOU S B, KONG J J, et al. Effects of applying amount of selenium on selenium accumulation and selenium forms of *Astragalus sinicus*[J]. Journal of Soil and Water Conservation, 2014, 28(4):304-309.

[34] ZHOU Z D, YE H, MA J L, et al. Study on drought resistance on grafted seedlings from *Camellia* Xianglin No. 11 clone[J]. Journal of Central South University of Forestry and Technology, 2015, 35(2):54-58.

Chapter V Intercropping

1 Introduction

Selenium (Se)is an essential trace element in human body. Humans obtain less than 40 μg of Se a day from the diet, and Se poisoning occurs after intake of concentration above 400 μg a day (Supriatin et al., 2015; El-Ramady et a., 2015). Se deficiency is associated with various human diseases such as heart disease, cancer, and reproductive disorders. High intake of Se for humans leads to loss of hair and nails and damage to the nervous and digestive systems (El-Ramady et al., 2015). The human body can not synthesize Se so it is obtained through diet, while plant Se is the primary source of Se in human body (Rayman, 2008; Winkel et al., 2012). Application of Se fertilization is a very common method for improving plant Se content, but this approach is expensive with high environmental pollution risk (Zhang, 2018). Therefore, exploring effective and eco-friendly methods to improve Se accumulation in crops is imperative.

Intercropping improves the rate of utilization of light, temperature, water, fertilizer and other resources by crops. Moreover, intercropping modulates crop root exudates, soil pH value, and soil enzyme activity, ultimately improving crop yield and quality (Winkel et al., 2012). Intercropping of maize with soybean is a typical representative of rational utilization of various resources for the production of crops (Fu et al., 2019). Intercropping maize with legumes improved the absorption of nutrients and increases the yield of the two crops (Li et al., 2008). Intercropping the hyperaccumulator *Thlaspi caerulescens* with the non-hyperaccumulator *Thlaspi arvense* under heavy metal contaminated condition promotes the growth of the two plants, increases zinc (Zn)accumulation in *T. caerulescens*, and decreases Zn uptake in *T. arvense* (Whiting et al., 2001). Intercropping cadmium (Cd)-hyperaccumulator *Solanum photeinocarpum* or its post-grafting generations with loquat seedlings promotes Cd accumulation and growth of the two plants (Liu et al., 2022). In addition, intercropping of hyperaccumulator *Galinsoga parviflora* with the accumulator plants *Capsella bursa-pastoris*, *Cardamine hirsuta*, and *Galium aparine* exhibites different effects on Cd accumulation and growth of these plants, including promotion, inhibition, or no effects (Lin et al., 2014). Three varieties (red, green and black) of eggplant seedlings intercropped under Se-rich soil conditions show increased Se accumulation and growth rate. Moreover, the three varieties exhibited increased photosynthetic pigment content, antioxidant enzyme activities, and soluble protein content (Pan et al., 2021). Intercropping of three radish genotypes has various effects on their growth, physiology, and Se accumulation. Some intercropping combinations of radish promote growth and Se accumulation of radish and improve their resistances to Se tress, whereas other intercropping combinations show the opposite effects (Tang, 2018). Intercropping of three genotypes of cherry tomato has various effects on the yield and quality of fruits, but all combinations show increased Se content in various organs (Liang et al., 2021). These findings indicate that different intercropping combinations of crops may produce varying effects on their growth and Se accumulation, implying that appropriate intercropping combinations can promote Se accumulation in crops. Studies should explore suitable intercropping combinations to improve Se accumulation in crops.

Cyphomandra betacea is a perennial evergreen fruit tree with high edible and ornamental values (Dong et al., 2022; Zhang et al., 2005). The Se accumulation capacity of *C. betacea* is lower than that of other Se-rich vegetables and fruits (Lin et al., 2020a). *Solanum photeinocarpum* and *Solanum nigrum* are annual to perennial *Solanum* sect. *Solanum* wild vegetables with a high Se accumulation capacity (Wu et al., 2018; Huang et al., 2018; Wang et al., 2019b). *S. photeinocarpum* is a diploid plant, *S. nigrum* is a hexaploid plant, whereas *S. nigrum* was obtained from the breeding of *S. photeinocarpum* (Xu et al., 2004). Intercropping *C. betacea* with *S. photeinocarpum* and *S. nigrum* may improve the Se accumulation capacities of these plants. Studies have not explored the effect of intercropping these plants on the Se accumulation capacities. Therefore, in this study, *C. betacea* was intercropped with *S. photeinocarpum* (diploid), colchicine-induced *S. photeinocarpum* (tetraploid), and *S. nigrum* (hexaploid), and the Se uptake by these plants was evaluated to determine the best combination of sect. *Solanum* and *C. betacea* for improving Se accumulation in these plants.

2 Materials and methods

2.1 Materials

The *C. betacea* seeds were collected from a five-year-old fruiting tree growing at the Chengdu Campus of Sichuan Agricultural University (30°42′N, 103°51′E) in September 2021. The seeds were air-dried. The seeds of *S. photeinocarpum* (diploid) and *S. nigrum* (hexaploid) were obtained from the farmland around the Chengdu Campus in June 2021. The shoot tips of the diploid seedlings at the stage of unfolded cotyledons were treated with colchicine (0.3%) for 48 h, and the tetraploid was induced. Tetraploid seeds were obtained from the diploid plants treated with colchicine in September 2021. Diploid, tetraploid, and hexaploid seeds were air-dried. The seeds of four plants were planted in seedling trays filled with wet perlite in February 2022. Hoagland nutrient solutions were used to irrigate the seedlings every three days after emergence.

The soil used in this experiment was collected from the farmland around the Chengdu Campus. The soil species was fluvo-aqic, and the basic phyicochemical properties of soil samples are presented in Table 5-1.

Table 5-1 The basic chemical properties of soil samples

Soil type	pH value	Organic matter/(g/kg)	Total N/(g/kg)	Total P/(g/kg)	Total K/(g/kg)	Alkaline hydrolyzed N/(mg/kg)	Available P/(mg/kg)	Available K/(mg/kg)	Total Se/(mg/kg)
Fluvo-aquic	7.71	15.29	1.85	11.88	15.38	87.99	55.78	41.96	0.12

2.2 Experimental design

The experiment was conducted in a greenhouse at the Chengdu Campus. The soil was treated in March 2022 according to Lin et al. (2020b). 3.0 kg soil was placed in a plastic pot (21 cm diameter and 20 cm depth), and Na_2SeO_3 (analytical grade) was added to the soil to obtain 5 mg/kg Se concentration (Lin et al., 2022). Then, the soil was watered to maintain the soil moisture content at 80% of the field capacity for one month. In April 2022, uniform seedlings of plant seedlings were transplanted in the pot. In each pot, four seedlings were planted with even distribution in all four directions. The experiment comprised seven treatments: *C. betacea* monoculture, diploid monoculture, tetraploid monoculture, hexaploid monoculture, *C. betacea* intercropped with the diploid plant, *C. betacea* intercropped with the tetraploid plant, and *C. betacea* intercropped with the hexaploidy plant. Four seedlings of one plant species were planted in each pot for the monoculture treatments. Two seedlings of *C. betacea* and two seedlings each of diploid, tetraploid, or hexaploid were planted together in each pot for the intercropping treatments. Each treatment was conducted in triplicate (three pots) and a completely randomized design was used in the study. The plants were irrigated using tap water.

2.3 Determination of indicators

Two months after plants transplanting, the third mature leaves of plants from top were collected to determine various parameters, including the content of photosynthetic pigments (chlorophyll a, chlorophyll b, and carotenoid), activities of antioxidant enzyme [superoxide dismutase (SOD), peroxidase (POD), and catalase (CAT)], and soluble protein content following the methods reported by Lin et al. (2020b) and Hao et al. (2004). Subsequently, the plants were harvested and treated as described by Li et al. (2022). The dry weights (biomass) of roots and shoots were measured using an electronic balance. The plant samples were dried, ground, and digested with nitric acid and perchloric acid. Furthermore, the digestion solutions were reduced with hydrochloric acid and the Se concentration was determined using the hydride generation-atomic fluorescence spectrometer (AFS-9700, Beijing Haiguang Instrument Co., Ltd., Beijing, China) (Li et al., 2022). The translocation factor (TF, Se content in shoot/root) was calculated as described previously (Rastmanesh et al., 2010).

2.4 Statistical analyses

The data was analyzed using SPSS 27.0 software (IBM, Chicago, IL, USA). Data was normalized and subjected to a homogeneity test. One-way analysis of variance was then conducted for comparison of the multiple groups, followed by Duncan's multiple range test for pairwise comparison ($P < 0.05$). Relationships of the biomass and Se content between the monoculture and intercropping treatments were evaluated using regession analysis. Pearson's correlation analysis was conducted to determine the correlations among the different indicators of each plant. Grey relational analysis was performed to explore the relationships of the different indicators with the shoot Se content of each plant as described by Wang (2019a) and Ma et al. (2022).

3 Results

3.1 Biomass

The root and shoot biomass of *C. betacea* was lower than that of the diploid, tetraploid, and hexaploid plants. The biomass order in different ploidies was diploid < tetraploid < hexaploid [Figure 5-1(a) and Figure 5-1(b)]. Compared with *C. betacea* monoculture, intercropped with diploid, tetraploid, and hexaploid decreased the root biomass of *C. betacea* by 6.07%, 15.63%, and 19.23%, respectively, and decreased the shoot biomass of *C. betacea* by 9.30%, 20.15%, and 23.13%, respectively. The root and shoot biomass of different ploidies intercropped with *C. betacea* was higher compared with that of their respective monoculture. In addition, the root biomass of plants intercropped with diploid, tetraploid, and hexaploid plants was linearly positively correlated with the root biomass of the respective monoculture [Figure 5-2(a)]. The root biomass of intercropped *C. betacea* was negatively correlated with the root biomass of monoculture diploid, tetraploid, and hexaploid plants [Figure 5-2(b)]. The shoot biomass of intercropped diploid, tetraploid, and hexaploid plants was positively related to the shoot biomass of the corresponding monoculture [Figure 5-2(c)]. The shoot biomass of intercropped *C. betacea* was linearly negatively correlated with the shoot biomass of monoculture diploid, tetraploid, and hexaploid plants [Figure 5-2(d)].

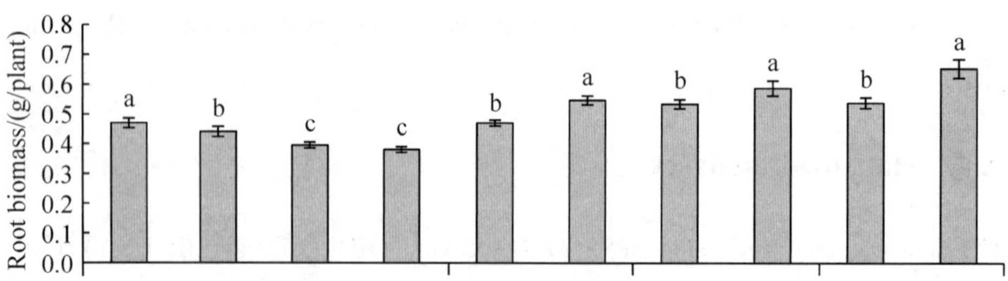

(a)

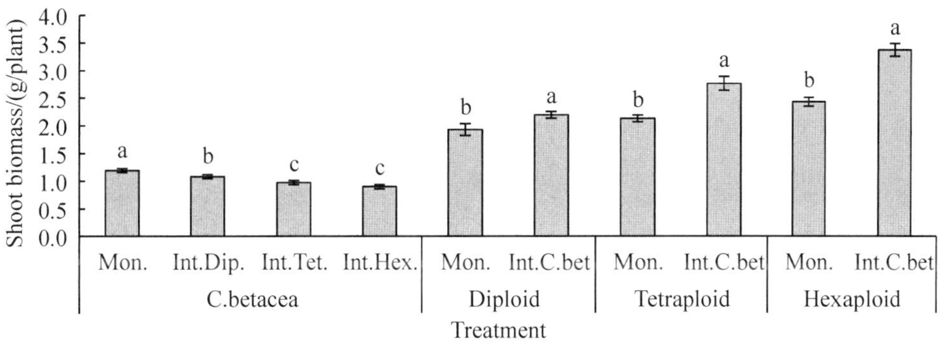

(b)

Note: Different lowercase letters indicate significant differences among the treatments (Duncan's multiple range test, $P < 0.05$). Mon. = monoculture; Int. Dip. = intercropped with diploid; Int.Tet. = intercropped with tetraploid; Int.Hex. = intercropped with hexaploid; Int.C.bet = intercropped with *C. betacea*.

Figure 5-1 Root biomass (a) and shoot biomass (b) of plants

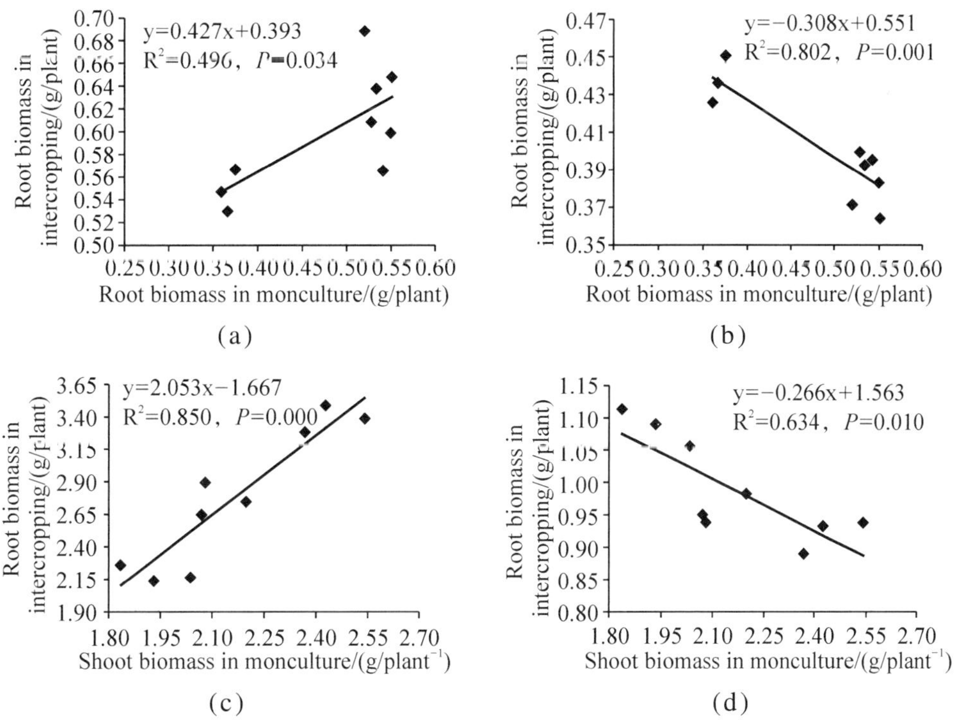

Note: (a) Root biomass of the diploid, tetraploid, and hexaploid in intercropping plotted against their root biomass in monoculture; (b) root biomass of *C. betacea* in intercropping plotted against root biomass of the diploid, tetraploid, and hexaploid in monoculture; (c) shoot biomass of the diploid, tetraploid, and hexaploid in intercropping plotted against their shoot biomass in monoculture; (d) shoot biomasses of *C. betacea* in intercropping plotted against shoot biomass of the diploid, tetraploid, and hexaploid in monoculture.

Figure 5-2 Linear regression relationships of plants' biomass

3.2 Photosynthetic pigment content

C. betacea intercropped with the diploid, tetraploid, and hexaploid plant showed lower content of chlorophyll a, chlorophyll b, and carotenoid compared with the levels in *C. betacea* monoculture (Table 5-2). Notably, the content of chlorophyll a/b of *C. betacea* monoculture were not significantly different compared with *C. betacea* intercropped with the diploid, tetraploid, and hexaploid plants. Diploid intercropped with *C. betacea* showed no significant differences in the content of chlorophyll a, carotenoid relatively to that of the monoculture of diploid. The content of chlorophyll b was higher in the diploid intercropped with *C. betacea*, and lower chlorophyll a/b compared with that of monoculture diploid. Tetraploid intercropped with *C. betacea* showed higher content of chlorophyll a, chlorophyll b, and carotenoid relatively to that of tetraploid monoculture, whereas there was no significant in chlorophyll a/b content between tetraploid monoculture and intrcropped tetraploid. Hexaploi intercropped with *C. betacea* exhibited higer levels of chlorophyll a and carotenoid compared with that in hexaploid monoculturs. On the contrary, chlorophyll b content and chlorophyll a/b were not significantly different between hexaploid monoculture and intercropped hexaploid.

Table 5-2 Photosynthetic pigment content in plants

Treatment	Chlorophyll a/(mg/g)	Chlorophyll b/(mg/g)	Chlorophyll a/b	Carotenoid/(mg/g)
C. betacea				
Mono.	1.014±0.047a	0.445±0.015a	2.281±0.147a	0.174±0.002a
Inter. diploid	0.896±0.031b	0.399±0.012b	2.243±0.018a	0.155±0.004b
Inter. tetraploid	0.790±0.022c	0.359±0.014c	2.203±0.059a	0.140±0.008c
Inter. hexaploid	0.789±0.018c	0.348±0.011c	2.265±0.023a	0.135±0.006c
Diploid				
Mono.	1.668±0.052a	0.680±0.030b	2.454±0.059a	0.285±0.012a
Inter. *C. betacea*	1.768±0.086a	0.773±0.035a	2.288±0.051b	0.302±0.013a
Tetraploid				
Mono.	1.873±0.055b	0.791±0.045b	2.372±0.078a	0.309±0.005b
Inter. *C. betacea*	2.056±0.067a	0.872±0.018a	2.358±0.097a	0.336±0.011a
Hexaploid				
Mono.	1.976±0.094b	0.857±0.039a	2.308±0.144a	0.322±0.016b
Inter. *C. betacea*	2.154±0.061a	0.876±0.023a	2.459±0.026a	0.355±0.010a

Note: Different lowercase letters indicate significant differences among the treatments (Duncan's multiple range test, $P < 0.05$). Mono. = monoculture; Inter.diploid = intercropped with diploid; Inter. tetraploid = intercropped with tetraploid; Inter. hexaploid = intercropped with hexaploid; Inter. *C. betacea* = intercropped with *C. betacea*.

3.3 Antioxidantenzyme activities and soluble protein content

C. betacea intercropped with diploid and tetraploid did not exhibit significant differences in SOD activity compared with *C. betacea* monoculture (Table 5-3). *C. betacea* intercropped with hexaploid showed lower SOD activity than *C. betacea* monoculture. *C. betacea* intercropped with the diploid, tetraploid, and hexaploid showed higher POD activity, CAT activity, and soluble protein content compared with monoculture of the *C. betacea*. Diploid intercropped with *C. betacea* showed higher SOD activity, POD activity, CAT activity and soluble protein content compared with the monoculture of diploid. Tetraploid intercropped with *C. betacea* did not exhibit significant difference in SOD activity compared with the tetraploid monoculture. However, tetraploid intercropped with *C. betacea* had higher POD activity, CAT activity and soluble protein content relative to the tetrploid monoculture. Hxaploid intercropped with *C. betacea* did not exhibit significant diferences in SOD activity and CAT activity relatively to hexaploidy monoculture. Compared with the hexaploid monoculture, hexaploid intercropped with *C. betacea* had higher POD activity and soluble protein content.

Table 5-3 Antioxidant enzyme activities and soluble protein content

Treatment	SOD activity/(U/g)	POD activity/[U/(g/min)]	CAT activity/[mg/(g/min)]	Soluble protein/(mg/g)
C. betacea				
Mono.	361. 1±16. 43a	3,889±48. 82c	5. 876±0. 202c	12. 21±0. 56c
Inter. diploid	353. 1±17. 71ab	4,397±91. 16b	6. 459±0. 131b	13. 26±0. 47b
Inter. tetraploid	332. 7±10. 01ab	5,410±79. 98a	6. 585±0. 185b	13. 91±0. 52ab
Inter. hexaploid	327. 6±13. 71b	5,426±58. 01a	7. 132±0. 191a	14. 56±0. 49a
Diploid				
Mono.	235. 1±8. 67b	4,666±87. 25b	6. 813±0. 089b	10. 70±0. 20b
Inter. *C. betacea*	256. 8±9. 49a	5,771±157. 9a	7. 064±0. 078a	12. 99±0. 58a
Tetraploid				
Mono.	258. 4±9. 22a	5,307±102. 1b	7. 112±0. 095b	12. 18±0. 24b
Inter. *C. betacea*	268. 1±5. 44a	7,334±316. 9a	7. 288±0. 049a	15. 94±0. 14a
Hexaploid				

continued table

Treatment	SOD activity/ (U/g)	POD activity/ [U/(g/min)]	CAT activity/ [mg/(g/min)]	Soluble protein/ (mg/g)
Mono.	272. 3±7. 78a	6,811±225. 3b	7. 484±0. 035a	13. 52±0. 49b
Inter. *C. betacea*	283. 5±6. 54a	8,694±232. 2a	7. 542±0. 054a	19. 58±0. 42a

Note: Different lowercase letters indicate significant differences among the treatments (Duncan's multiple range test, $P < 0.05$). Mono. = monoculture; Inter. diploid = intercropped with diploid; Inter. tetraploid = intercropped with tetraploid; Inter. hexaploid = intercropped with hexaploid; Inter. *C. betacea* = intercropped with *C. betacea*.

3.4 Se Content and translocation factor

C. betacea intercropped with the diploid, tetraploid, and hexaploid exhibited higher root and shoot Se content compared with that of their respective monocultures [Figure 5-3(a) and Figure 5-3(b)]. *C. betacea* intercropped with the diploid, tetraploid, and hexaploid showed higher Se content in shoots by 13. 73%, 17. 49%, and 26. 50%, respectively, compared with *C. betacea* monoculture. Intercropping of different ploidies with *C. betace* a increased the shoot Se content in diploid, tetraploid, and hexaploid plants by 35. 22%, 68. 86%, and 74. 46%, respectively, compared with the diploid, tetraploid, and hexaploid monocultures. In addition, the root Se content in *C. betacea* intercropped with the diploid, tetraploid, and hexaploid were positively correlated with the root Se content of the corresponding monoculture [r = 0. 760, N = 9, P = 0. 018, Figure 5-4(a)]. Root Se content in *C. betacea* intercropped with different ploidies were positively correlated with the root Se content in corresponding diploid, tetraploid, and hexaploid monocultures [r = 0. 739, N = 9, P = 0. 023, Figure 5-4(b)]. The shoot Se content in *C. betacea* intercropped with diploid, tetraploid, and hexaploid plants were linearly positively correlated with shoot Se content in the corresponding monoculture [r = 0. 927, N = 9, P = 0. 000, Figure 5-4(c)]. The shoot Se content in *C. betacea* intercropped with different ploidies were linearly positively correlated with the shoot Se content in diploid, tetraploid, and hexaploid monocultures [r = 0. 958, N = 9, P = 0. 000, Figure 5-4(d)].

The findings showed that *C. betacea* intercropped with the diploid had higher TF relatively to the *C. betacea* monocultur. *C. betacea* intercropped with the tetraploid showed no significant difference in TF, whereas *C. betacea* intercropped with the hexaploid had lower TF compared with the *C. betacea* monoculture [Figure 5-3 (c)]. Diploid, tetraploid, and hexaploid plants intercropped with *C. betacea* exhibited higher TFs compared with the TFs in the diploid, tetraploid, and hexaploidy monocultures, respectively.

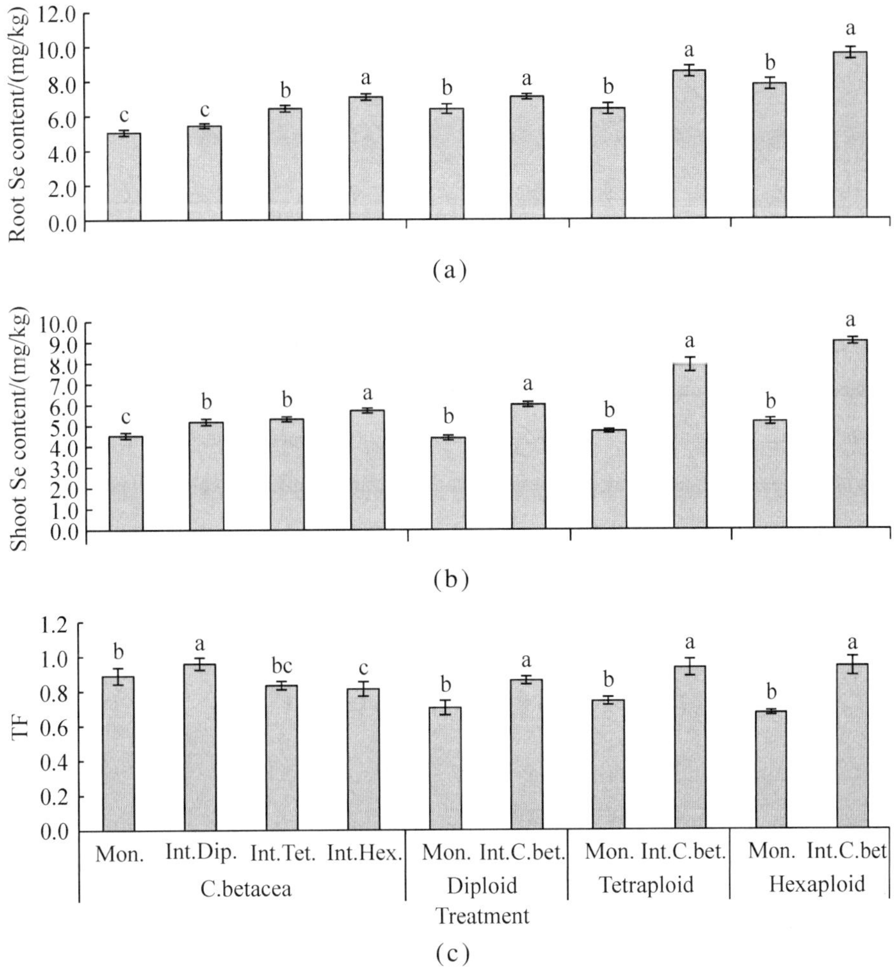

Note: Different lowercase letters indicate significant differences among the treatments (Duncan's multiple range test, $P < 0.05$). Mon. = monoculture; Int. Dip. = intercropped with diploid; Int. Tet. = intercropped with tetraploid; Int. Hex. = intercropped with hexaploid; Int. C. bet = intercropped with *C. betacea*.

Figure 5-3 Root Se cotent (a), shoot Se content (b), and TF (c) of plants

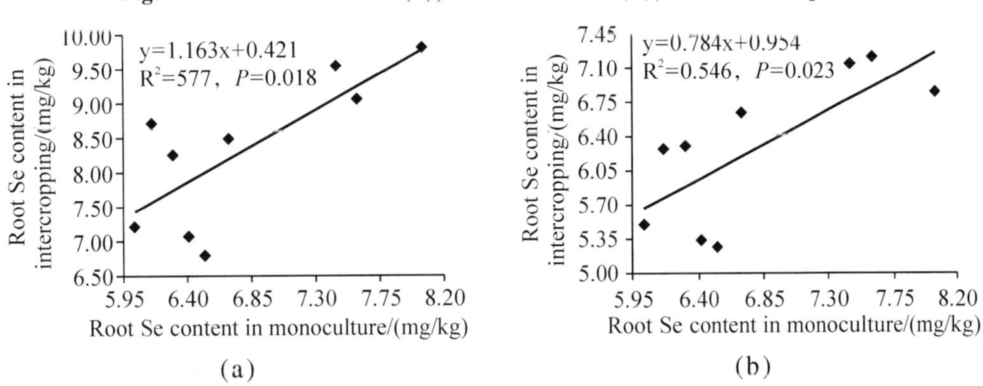

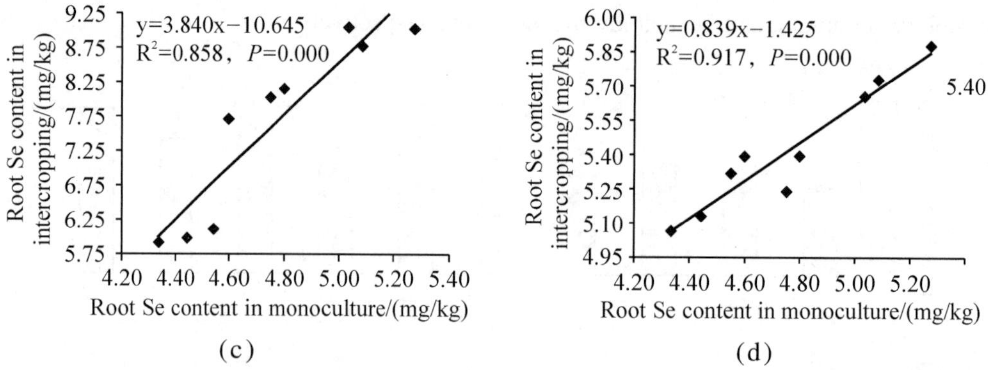

Note: (a) Root Se content in the diploid tetraploid, and hexaploid in intercropping plotted against their root Se content in monoculture; (b) root Se content in *C. betacea* in intercropping plotted against root Se content in the diploid, tetraploid, and hexaploid in monoculture; (c) shoot Se content in the diploid, tetraploid, and hexaploid in intercropping plotted against their shoot Se content in monoculture; (d) shoot Se content in *C. betacea* in intercropping plotted against shoot Se content in the diploid, tetraploid, and hexaploid in monoculture.

Figure 5-4 Linear regression relationships of Se content in plants

3.5 Relationship of Different Parameters in Plants

The root biomass and shoot biomass of *C. betacea* were highly significantly ($P<0.01$) positively correlated with the chlorophyll a content, chlorophyll b content, carotenoid content, and SOD activity (Table 5-4). The root biomass and shoot biomass were highly significantly ($P<0.01$) negatively correlated with the POD activity, CAT activity, and soluble protein content. The root Se content and shoot Se content were highly significantly ($P<0.01$) positively correlated with the POD activity, CAT activity, and soluble protein content, and were highly significantly ($P<0.01$) or significantly ($0.01\leqslant P<0.05$) negatively correlated with the root biomass, shoot biomass, chlorophyll a content, chlorophyll b content, carotenoid content, and SOD activity. The root Se content was highly significantly ($P<0.01$) positively correlated with the shoot Se content.

The root biomass of diploid plant was highly significantly ($P<0.01$) or significantly ($0.01\leqslant P<0.05$) positively correlated with the POD activity, CAT activity, soluble protein content, and TF (Table 5-5). The shoot biomass was highly significantly ($P<0.01$) or significantly ($0.01\leqslant P<0.05$) positively correlated with the CAT activity, soluble protein content, and TF. The root Se content was highly significantly ($P<0.01$) or significantly ($0.01\leqslant P<0.05$) positively correlated with the root biomass, carotenoid content, POD activity, and soluble protein content. The shoot Se content was highly significantly ($P<0.01$) or significantly ($0.01 \leqslant P < 0.05$) positively correlated with the root biomass, shoot biomass, chlorophyll b content, POD activity, CAT activity, soluble protein content, and TF, and was significantly ($0.01 \leqslant P < 0.05$) negatively correlated with the content of chlorophyll a/b. The root Se content was highly significantly ($P < 0.01$) positively correlated with the shoot Se content.

The root biomass of tetraploid plant was significantly ($0.01 \leqslant P < 0.05$) positively correlated with the chlorophyll a content, chlorophyll b content, carotenoid content, CAT activity, and soluble protein content (Table 5-6). The shoot biomass was highly significantly ($P < 0.01$) or significantly ($0.01 \leqslant P < 0.05$) positively correlated with the chlorophyll b content, POD activity, soluble protein content, and TF. The root Se content and shoot Se content were highly significantly ($P < 0.01$) or significantly ($0.01 \leqslant P < 0.05$) positively correlated with the root biomass, shoot biomass, chlorophyll a content, chlorophyll b content, carotenoid content, POD activity, CAT activity, soluble protein content, and TF. The root Se content was highly significantly ($P < 0.01$) positively correlated with the shoot Se content.

The root biomass of hexaploid plant was highly significantly ($P < 0.01$) or significantly ($0.01 \leqslant P < 0.05$) positively correlated with the chlorophyll a content, SOD activity, POD activity, soluble protein content, and TF (Table 5-7). The shoot biomass was highly significantly ($P < 0.01$) positively correlated with the POD activity, soluble protein content, and TF. The root Se content was highly significantly ($P < 0.01$) or significantly ($0.01 \leqslant P < 0.05$) positively correlated with the root biomass, shoot biomass, carotenoidcontent, POD activity, and soluble protein content, and TF. The shoot Se content was highly significantly ($P < 0.01$) or significantly ($0.01 \leqslant P < 0.05$) positively correlated with the root biomass, shoot biomass, chlorophyll a content, carotenoid content, POD activity, soluble protein content, and TF. The root Se content was highly significantly ($P < 0.01$) positively correlated with the shoot Se content.

Table 5-4 Correlations among the different indicators of *C. betacea*

Indicator	Root biomass	Shoot biomass	Chlorophyll a content	Chlorophyll b content	Chlorophyll a/b	Carotenoid content	SOD activity	POD activity	CAT activity	Soluble protein content	Root Se content	Shoot Se content	TF
Root biomass													
Shoot biomass	0. 922 * *												
Chlorophyll a content	0. 880 * *	0. 953 * *											
Chlorophyll b content	0. 923 * *	0. 926 * *	0. 949 * *										
Chlorophyll a/b	0. 094	0. 315	0. 397	0. 090									
Carotenoid content	0. 893 * *	0. 942 * *	0. 918 * *	0. 952 * *	0. 129								
SOD activity	0. 709 * *	0. 776 * *	0. 649 *	0. 747 * *	−0. 107	0. 787 * *							
POD activity	−0. 956 * *	−0. 957 * *	−0. 928 * *	−0. 935 * *	−0. 217	−0. 923 * *	−0. 738 * *						
CAT activity	−0. 874 * *	−0. 838 * *	−0. 824 * *	−0. 912 * *	0. 061	−0. 914 * *	−0. 685 *	0. 833 * *					
Soluble protein content	−0. 779 * *	−0. 894 * *	−0. 804 * *	−0. 830 * *	−0. 110	−0. 858 * *	−0. 759 * *	0. 783 * *	0. 749 * *				
Root Se content	−0. 912 * *	−0. 930 * *	−0. 866 * *	−0. 865 * *	−0. 208	−. 841 * *	−0. 694 *	0. 933 * *	0. 823 * *	0. 794 * *			
Shoot Se content	−0. 922 * *	−0. 930 * *	−0. 898 * *	−0. 931 * *	−0. 127	−0. 916 * *	−0. 733 * *	0. 881 * *	0. 936 * *	0. 855 * *	0. 870 * *		
TF	0. 128	0. 231	0. 254	0. 145	0. 403	0. 246	0. 042	−0. 236	−0. 141	−0. 082	−0. 190	−0. 212	

Note: * * = Correlation is significant at the 0. 01 level (2-tailed test). * = Correlation is significant at the 0. 01 level (2-tailed test). N = 12.

Table 5-5 Correlations among the different indicators of the diploid

Indicator	Root biomass	Shoot biomass	Chlorophyll a content	Chlorophyll b content	Chlorophyll a/b	Carotenoid content	SOD activity	POD activity	CAT activity	Soluble protein content	Root Se content	Shoot Se content	TF
Root biomass													
Shoot biomass	0.813 *												
Chlorophyll a content	0.691	0.531											
Chlorophyll b content	0.810	0.739	0.892 *										
Chlorophyll a/b	−0.733	−0.779	−0.537	−0.859 *									
Carotenoid content	0.705	0.322	0.662	0.706	−0.592								
SOD activity	0.691	0.672	0.332	0.655	−0.827 *	0.315							
POD activity	0.976 * *	0.784	0.692	0.869 *	−0.841 *	0.765	0.772						
CAT activity	0.909 *	0.883 *	0.422	0.582	−0.609	0.401	0.551	0.836 *					
Soluble protein content	0.939 * *	0.813 *	0.830 *	0.942 * *	−0.816 *	0.662	0.759	0.954 * *	0.733				
Root Se content	0.888 *	0.552	0.61	0.772	−0.767	0.881 *	0.699	0.945 * *	0.675	0.840 *			
Shoot Se content	0.961 * *	0.864 *	0.727	0.913 *	−0.883 *	0.692	0.788	0.984 * *	0.842 *	0.975 * *	0.881 *		
TF	0.896 *	0.955 * *	0.711	0.893 *	−0.860 *	0.501	0.759	0.899 *	0.849 *	0.945 * *	0.713	0.960 * *	

Note: * * =Correlation is significant at the 0.01 level (2-tailed test). * = Correlation is significant at the 0.01 level (2-tailed test). N = 6.

Table 5-6 Correlations among the different indicators of the tetraploid

Indicator	Root biomass	Shoot biomass	Chlorophyll a content	Chlorophyll b content	Chlorophyll a/b	Carotenoid content	SOD activity	POD activity	CAT activity	Soluble protein content	Root Se content	Shoot Se content	TF
Root biomass													
Shoot biomass	0.876 *												
Chlorophyll a content	0.823 *	0.784											
Chlorophyll b content	0.902 *	0.867 *	0.854 *										
Chlorophyll a/b	-0.333	-0.326	0.038	-0.487									
Carotenoid content	0.874 *	0.788	0.963 * *	0.788	0.114								
SOD activity	0.563	0.613	0.308	0.298	-0.011	0.492							
POD activity	0.787	0.968 * *	0.803	0.769	-0.105	0.807	0.596						
CAT activity	0.865 *	0.739	0.946 * *	0.847 *	-0.037	0.929 * *	0.209	0.734					
Soluble protein content	0.835 *	0.958 * *	0.855 *	0.782	-0.043	0.881 *	0.646	0.987 * *	0.780				
Root Se content	0.937 * *	0.960 * *	0.880 *	.873 *	-0.178	0.907 *	0.566	0.945 * *	0.880 *	0.963 * *			
Shoot Se content	0.863 *	0.948 * *	0.913 *	0.832 *	-0.047	0.922 * *	0.571	0.972 * *	0.849 *	0.992 * *	0.975 * *		
TF	0.736	0.888 *	0.895 *	0.747	0.082	0.882 *	0.545	0.950 * *	0.766	0.969 * *	0.897 *	0.972 * *	

Note: * * =Correlation is significant at the 0.01 level (2-tailed test). * = Correlation is significant at the 0.01 level (2-tailed test). N = 6.

Table 5-7 Correlations among the different indicators of the hexaploid

Indicator	Root biomass	Shoot biomass	Chlorophyll a content	Chlorophyll b content	Chlorophyll a/b	Carotenoid content	SOD activity	POD activity	CAT activity	Soluble protein content	Root Se content	Shoot Se content	TF
Root biomass													
Shoot biomass	0.932 * *												
Chlorophyll a content	0.870 *	0.805											
Chlorophyll b content	0.534	0.359	0.463										
Chlorophyll a/b	0.597	0.646	0.793	−0.172									
Carotenoid content	0.770	0.764	0.484	−0.008	0.540								
SOD activity	0.815 *	0.683	0.972 * *	0.598	0.669	0.345							
POD activity	0.891 *	0.968 * *	0.726	0.155	0.699	0.883 *	0.572						
CAT activity	0.456	0.586	0.312	−0.377	0.607	0.756	0.123	0.723					
Soluble protein content	0.957 * *	0.991 * *	0.803	0.374	0.634	0.810	0.691	0.969 * *	0.626				
Root Se content	0.939 * *	0.893 *	0784	0.234	0.707	0.921 * *	0.681	0.939 * *	0.656	0.929 * *			
Shoot Se content	0.957 * *	0.983 * *	0.824 *	0.317	0.695	0.839 *	0.706	0.980 * *	0.597	0.985 * *	0.956 * *		
TF	0.921 * *	0.994 * *	0.811	0.337	0.668	0.754	0.686	0.964 * *	0.543	0.975 * *	0.886 *	0.983 * *	

Note: * * = Correlation is significant at the 0.01 level (2-tailed test). * = Correlation is significant at the 0.01 level (2-tailed test). N = 6.

3.6 Grey Relational Analysis

The grey relationships of different indicators with the shoot Se content were evaluated and the results showed that all indicators were correlated with the shoot Se content (Figure 5-5). The top four indicators of *C. betacea* with the highest grey correlation coefficients were CAT activity, POD activity, soluble protein content, and root Se content [Figure 5-5(a)]. The top indicators wih the highest grey correlation coefficients in diploid plant were POD activity, soluble protein content, TF, and root biomass [Figure 5-5(b)]. The top indicators for tetraploid plant were soluble protein content, POD activity, root Se content, and TF [Figure 5-5(c)], and the top indicators for hexaploid plant were soluble protein content, shoot biomass, TF, and POD activity [Figure 5-5(d)].

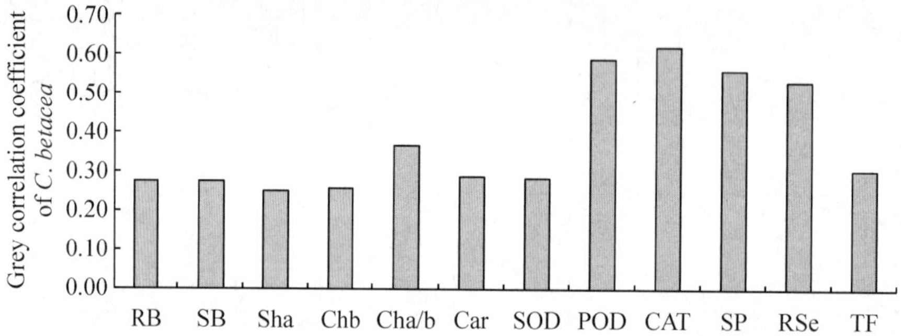

(a)

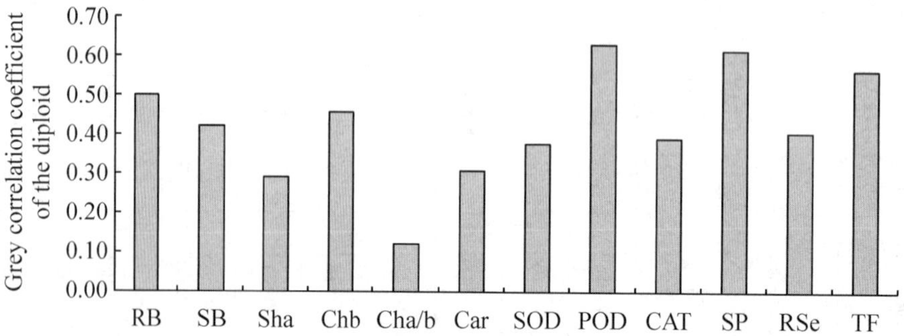

(b)

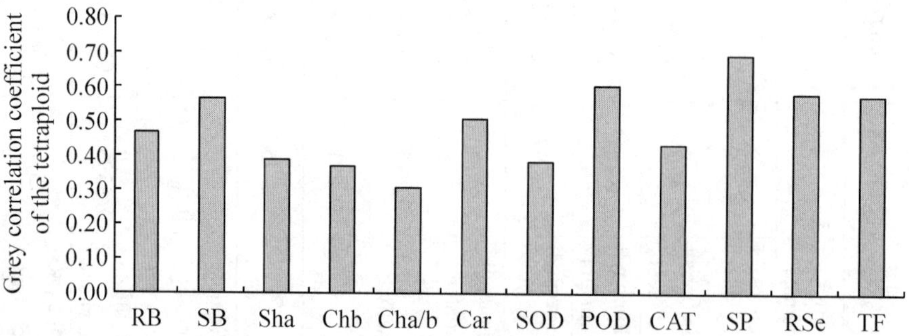

(c)

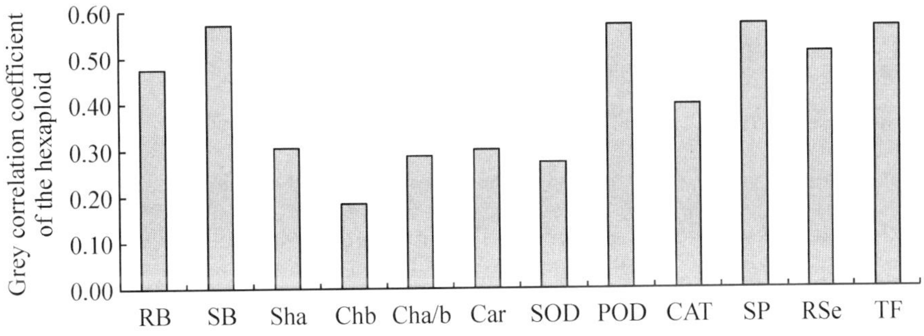

(d)

Note: RB = root biomass; SB = shoot biomass; Cha = chlorophyll a content; Chb = chlorophyll b content; Ch a/b = chlorophyll a/b; Car = carotenoid content; SOD = SOD activity; POD = POD activity; CAT = CAT activity; SP = soluble protein content; RSe = root Se content; TF = translocation factor.

(a) *C. betacea*; (b) diploid; (c) tetraploid; (d) hexaploid.

Figure 5-6 Grey correlation coefficients of different indicators with the shoot Se content

4 Discussion

Root exudates of intercropped plants diffuse into the rhizosphere soil resulting in a change in the physical and chemical properties of rhizosphere soil. Intercropping changes soil organic acid content, soil pH value, soil nutrient availability, and soil enzyme activity, ultimately affecting the nutrient uptake and growth of plants (Liang et al., 2003). In addition, intercropping system results in competition between plants, which may inhibit the growth of the plants (Tang et al., 2017). Different species of intercropped eggplants exhibit an increase in the biomass of their seedlings under Se-rich soil conditions (Pan et al., 2017), whereas intercropping of white radish with the green radish results in an increase in the biomass of white radish and a decrease in the biomass of green radish (Tang, 2018). In the current study, *C. betacea* intercropped with sect. *Solanum* showed a decrease in the biomass of *C. betacea* and an increase in the biomass of sect. *Solanum*. This result is consistent with the findings from previous studies (Tang, 2018; Liang et al., 2021), but different from results from some studies (Whiting et al., 2001; Liu et al., 2022; Pan et al., 2021). This finding indicates competition for growth resources between *C. betacea* and sect. *Solanum*. The biomass of sect. *Solanum* was higher than that of *C. betacea* in the present study. Sect. *Solanum* are herbs, and their growth rate are higher than that of the seedlings of *C. betacea*, a woody plant. Therefore, sect. *Solanum* had a faster growth than *C. betacea*, inhibiting the growth of *C. betacea*. The biomass order of sect. *Solanum* was diploid < tetraploid <

hexaploidy. Intercropping of *C. betacea* with sect. *Solanum* showed that the biomass order was diploid < tetraploid < hexaploid. This finding implies that sect. *Solanum* competed and inhibited the growth of *C. betacea*. The biomass of intercropped sect. *Solanum* showed a positive linear relationship with the respective monoculture. The biomass of intercropped *C. betacea* exhibited a linear negative relationship with that of the sect. *Solanum* monocultures. This result further indicates the growth competition between *C. betacea* and sect. *Solanum*. Further studies should be conducted to explore the competition mechanism between the two species.

The content of photosynthetic pigments in plants reflects the photosynthetic capacity (Huang et al., 2020). Eggplant intercropped with the cadmium-hyperaccumulator *S. nigrum* exhibits a decrease in the photosynthetic pigment content in cadmium-contaminated soil. Tomato intercropped with *S. nigrum* under the same condition shows an increase in the photosynthetic pigment content (Huang et al., 2020). Grape seedlings intercropped with *S. nigrum* planted in cadmium contamination soil show increased chlorophyll a and total chlorophyll content (Ashraf et al., 2013). Maize intercropped with peanut and maize intercropped with soybean show increased chlorophyll content (Fu et al., 2019; Li et al., 2008.) In the current study, intercropping of *C. betacea* with sect. *Solanum* exhibited a decrease in the photosynthetic pigment content in *C. betacea* and an increase in the photosynthetic pigment content in sect. *Solanum*. These results indicate that the competition between *C. betacea* and sect. *Solanum* inhibits synthesis of photosynthetic pigments in *C. betacea*, and promotes the synthesis of photosynthetic pigments in sect. *Solanum*, which may be related to the competition for light, nutrients, and water resources.

High concentrations of Se in plants promote the production of reactive oxygen species (ROS) and induce oxidative stress in plants (Jiao et al., 2013). Plant defense mechanisms to alleviate Se stress include activating different antioxidant enzymes, such as SOD, POD and CAT (Nakamara et al., 2005; Elrashidi et al., 1987). Soluble protein is an essential osmotic regulator of plant cells and is implicated in plant stress resistance (Zhang et al., 2014). Eggplant intercropped with *S. nigrum* in cadmium-contaminated soil shows increased antioxidant enzyme activity and soluble protein content (Tang et al., 2018). Lettuce intercropped with the cadmium-hyperaccumulator *Galinsoga parviflora* exhibits increased antioxidant enzyme activity and soluble protein content (Longchamp et al., 2013). In the present study, intercropping of *C. betacea* with diploid and tetraploid plants had no significant effects on the SOD activity of *C. betacea*, whereas intercropping of *C. betacea* with hexaploid plants decreased SOD activity. Intercropping *C. betacea* with the diploid, tetraploid, and hexaploid increased the POD activity, CAT activity, and soluble protein content of *C. betacea*. Intercropping sect. *Solanum* with *C. betacea* increased the antioxidant enzyme activity and soluble protein content of sect. *Solanum*. These results indicate that intercropping of *C. betacea* with sect. *Solanum* can improve the resistance of the two plants to cadmium stress, consistent with findings from previous studies (Pan et al., 2021). Moreover, *C. betacea* intercropped with sect. *Solanum* may promote "rhizosphere dialogue", resulting in the transfer of the active compounds to each other to mutually improve their resistance to heavy metal stress (Inal et al., 2007).

The uptake of Se by plants is modulated by various factors such as soil pH value, soil redox conditions, and soil selenium state (Winkel et al., 2012). Se has a strong metalloid property and exists in inorganic forms with different oxidation states. The different soil oxidation states modulate the various valence states of Se (Jiang et al., 2009). The pathway for absorption of selenite by plants is similar to the absorption pathway for phosphate. Absorption of selenite by plants mainly occurs through the roots (Lin et al., 2020a; Lin et al., 2022). Intercropping improves the absorption of nutrients such as nitrogen, phosphorus, and potassium in maize and legumes (Fu et al., 2019; Li et al., 2008). Intercropping of heavy metal hyperaccumulator with non-hyperaccumulator increases heavy metal uptake in hyperaccumulator but decreases heavy metal uptake in non-hyperaccumulator (Whiting et al., 2001; Hao et al., 2004). Intercropping of different genotypes of tomato, eggplant, and radish with themselves exhibits a decrease or increase in Se uptake (Pan et al., 2021; Tang, 2018; Liang et al., 2021). In the current study, intercropping of *C. betacea* with sect. *Solanum* increased Se uptake in the two plants, consistent with findings on the other plants planted in cadmium-contaminated soils (Liu et al., 2022). These findings are consistent with results on the uptake level of Se by tomato, eggplant, and radish reported in previous studies (Liang et al., 2021). The Se content of intercropped sect. *Solanum* exhibited a positive linear relationship with the Se content of the monoculture. The Se content of intercropped *C. betacea* showed a linear negative relationship with the Se content in sect. *Solanum monocultures*. These results indicate that intercropping promoted Se uptake by *C. betacea* and sect. *Solanum* further confirming the "rhizosphere dialogue" (Inal et al., 2007) between *C. betacea* and sect. *Solanum*. Several active compounds may be exchanged between *C. betacea* and sect. *Solanum* to improve the Se accumulation capacity of the two plants. Intercropped of *C. betacea* with the diploid increased the TF of *C. betacea*, intercropping with the tetraploid had no significant effect on the TF of *C. betacea*, whereas intercropping with the hexaploid decreased the TF of *C. betacea*. However, intercropping of sect. *Solanum* with *C. betacea* increased the TFs of sect. *Solanum*. These results indicate that *C. betacea* intercropped with sect. *Solanum* had significant Se effects on sect. *Solanum* than *C. betacea*. This finding explains the higher percentage of shoot Se content in sect. *Solanum* than that in *C. betacea*. Correlation and grey relational analyses revealed that CAT activity, POD activity, and soluble protein content were the top three indicators significantly associated with *C. betacea* shoot Se content. In contrast, POD activity, soluble protein content, and TF were the top three indicators markedly associated with sect. *Solanum* shoot Se content. These results further imply that intercropping promoted Se absorption by increasing the resistance of the two plants to Se, and their action mechanisms should be explored further.

5 Conclusion

Intercropping of *C. betacea* with sect. *Solanum* inhibited the growth of *C. betacea* and promoted the growth of sect. *Solanum* as indicated by decrease or increase in their biomass and photosynthetic pigment content. In addition, intercropping *C. betacea* with sect. *Solanum* improved their resistances to cadmium by increasing the levels of antioxidant enzyme activities and soluble protein content in the two plants. Moreover, intercropping of the two plants increased the shoot Se content in *C. betacea* and sect. *Solanum*. The biomass and Se content of intercropped sect. *Solanum* exhibited linear relationships with that of the monocultures, and the biomass and Se content of intercropped *C. betacea* showed linear relationships with that of sect. *Solanum* monocultures. CAT activity, POD activity, and soluble protein content were the top three indicators highly associated with the Se content in *C. betacea* shoots. The POD activity, soluble protein content, and TF were the top three indicators significantly associated with the Se content in sect. *Solanum* shoots. Further studies should be conducted under field conditions to verify these findings. Furthermore, the effects of Se uptake by *C. betacea* fruits should be evaluated.

References

[1] ASHRAF M, HARRIS P J C. Photosynthesis under stressful environments: An overview[J]. Photosynthetica, 2013, 51(2):163-190.

[2] DONG Q, LI S M, GAO S J, et al. Comparison and comprehensive analysis the quality of *Cyphomandra betacea* from different provenance [J]. Food and Fermentation Industries, 2022, 48(4):266-273.

[3] EL-RAMADY H, ABDALLA N, ALSHAAL T, et al. Selenium in soils under climate change, implication for human health[J]. Environmental Chemistry Letters, 2015, 13 (1):1-19.

[4] ELRASHIDI M A, ADRIANO D C, WORKMAN S M, et al. Chemical equilibria of selenium in soils: A theoretical development[J]. Soil Science, 1987, 144(2):141-152.

[5] FU Z, ZHOU L, CHEN P, et al. Effects of maize-soybean relay intercropping on crop nutrient uptake and soil bacterial community [J]. Journal of Integrative Agriculture, 2019, 18(9):2006-2018.

[6] GARLAND G, BüNEMANN, E K, OBERSON A, et al. Plant-mediated rhizospheric interactions in maize-pigeon pea intercropping enhance soil aggregation and organic phosphorus storage[J]. Plant and Soil, 2017, 415(1-2):37-55.

[7] HAO Z B, CANG J, XU Z. Plant physiology experiment[M]. Harbin Institute of Technology Press Harbin, China, 2004.

[8] HUANG K W, LIN L J, AND LIAO M A. Study on selenium accumulation characteristics of *Solanum nigrum* [J]. IOP Conference Series: Earth and Environmental Science, 2018, 199(4):042027.

[9] HUANG K W, LIN L J, SHI J, et al. Effect of intercropping with grafting offspring of *Galinsoga parviflora* on cadmium accumulation of lettuce [J]. Soils, 2020, 52 (1): 127-133.

[10] INAL A, GUNES A, ZHANG F, et al. Peanut/maize intercropping induced changes in rhizosphere and nutrient concentrations in shoots[J]. Plant Physiology and Biochemistry, 2007, 45(5):350-356.

[11] JIANG C A, WU Q T, WU S H, et al. Effect of co-cropping*Sedum alfredii* with

different plants on metal uptake[J]. China Environmental Science, 2009, 29(9):985-990.

[12]JIAO N Y, YANG M K, NING T Y, et al. Effects of maize-peanut intercropping and phosphate fertilizer on photosynthetic characteristics and yield of intercropped peanut plants[J]. Chinese Journal of Plant Ecology, 2013, 37(11):1010-1017.

[13]LI N Y, LI Z A, DING Y Z, et al. Effects of intercropping different crops with maize on the Cd uptake by maize[J]. Journal of Applied Ecology, 2008, 19(6):1369-1373.

[14]LI Z, FAN R, PENG X, et al. Salicylic acid alleviates selenium stress and promotes selenium uptake of grapevine[J]. Physiology and Molecular Biology of Plants, 2022, 28(3):625-635.

[15]LIANG L, LIU J, LI X M, et al. Effects of intercropping with different genotypes of cherry tomato on fruit quality and selenium content[J]. Acta Agriculturae Zhejiangensis, 2021, 33(10):1870-1878.

[16]LIANG Y R, HU X H, ZHANG Y L, et al. Progress on physiological function research of plant peroxidase[J]. Journal of Inner Mongolia Agricultural University, 2003, 24(2):110-113.

[17]LIN L J, LUO L, YANG D Y, et al. Effects of intercropping with accumulator on cadmium accumulation of *Galinsoga parviflora*[J]. Journal of Soil and Water Conservation, 2014, 28(6):319-324.

[18] LIN L, SUN J, CUI T, et al. Selenium accumulation characteristics of *Cyphomandra betacea* (*Solanum betaceum*) seedlings[J]. Physiology and Molecular Biology of Plants, 2020a, 26(7):1375-1383.

[19]LIN L, WANG J, LIAO M, et al. *Artemisia argyi* water extract promotes selenium uptake of peach seedlings[J]. Frontiers in Plant Science, 2022, 13:1014454.

[20]LIN L, WU C, JIANG W, et al. Grafting increases cadmium accumulation in the post-grafting generations of the potential cadmium-hyperaccumulator *Solanum photeinocarpum*[J]. Chemistry and Ecology, 2020b, 36(7):685-704.

[21]LIU Y, LIU Q, XU X, et al. Effects of intercropping with *Solanum photeinocarpum* and its post-grafting generations on cadmium accumulation in loquat (*Eriobotrya japonica*)[J]. International Journal of Phytoremediation, 2022, 24(7):753-762.

[22]LONGCHAMP M, ANGELI N, CASTREC-ROUELLE M. Selenium uptake in *Zea mays* supplied with selenate or selenite under hydroponic conditions[J]. Plant and Soil, 2013, 362(1-2):107-117.

[23] MA Q, FASIH U H, MUHAMMAD F, et al. Selenium treated foliage and biochar treated soil for improved lettuce (*Lactuca sativa* L.) growth in Cd-polluted soil[J]. Journal of Cleaner Production, 2022, 335:130267.

[24] NAKAMARU Y, TAGAMI K, UCHIDA S. Distribution coefficient of selenium in Japanese agricultural soils[J]. Chemosphere, 2005, 58(10):1347-1354.

[25] PAN S, LU R, LI H, et al. Mutual intercropping affects selenium uptake of eggplant seedlings[J]. International Journal of Environmental Analytical Chemistry, 2021, 101(15):2866-2875.

[26] RASTMANESH F, MOORE F, AND KESHAVARZI B. Speciation and phytoavailability of heavy metals in contaminated soils in Sarcheshmeh area, Kerman Province, Iran[J]. Bulletin of Environmental Contamination and Toxicology, 2010, 85(5):515-519.

[27] RAYMAN M P. Food-chain selenium and human health: emphasis on intake[J]. The British Journal of Nutrition, 2008, 100(2):254-268.

[28] SUPRIATIN S, WENG L, COMANS R N J. Selenium speciation and extractability in Dutch agricultural soils[J]. Science of the Total Environment, 2015, 532:368-382.

[29] TANG F H. Effects of Intercropping on selenium enrichment of three genotypes radish[D]. Sichuan Agriculture University, Chengdu, China, 2018.

[30] TANG Y, HE J, YU X N, et al. Intercropping with*Solanum nigrum* and *Solanum photeinocarpum* from two ecoclimatic regions promotes growth and reduces cadmium uptake of eggplant seedlings[J]. Pedosphere, 2017, 27(3):638-644.

[31] WANG X. Application of grey relation analysis theory to choose high reliability of the network node[J]. Journal of Physics: Conference Series, 2019a, 1237(3):032056.

[32] WANG X, HUANG Z, LIN L, Liao M. Study on selenium accumulation characteristics of *Lycopersicon esculentum*, *Solanum melongena* and *Solanum nigrum*[J]. IOP Conference Series: Earth and Environmental Science, 2019b, 310(4):042065.

[33] WHITING S N, LEAKE J R, MCGRATH S P, et al. Hyperaccumulation of Zn by*Thlaspi caerulescens* can ameliorate Zn toxicity in the rhizosphere of cocropped *Thlaspi arvense*[J]. Environmental science and technology, 2001, 35(15):3237-3241.

[34] WINKEL L H, JOHNSON C A, LEN M, et al. Environmental selenium research: from microscopic processes to global understanding[J]. Environmental Science and Technology, 2012, 46(2):571-579.

[35]WU C F, BAO R F, LIAO H P, et al. Selenium accumulation characteristics of wild vegetable*Solanum photeinocarpum*[J]. Journal of Sichuan Agricultural University, 2018, 36(6):778-784.

[36]XU X F, ZHANG H Y, YUAN Q H, et al. Studies on chromosome of three types of *Solanum nigrum* in China[J]. Guihaia, 2004, 24(6):544-545.

[37]ZHANG C M. The regulatory effects of selenium nutrition in soil-rice system and cost-effectiveness analysis[D]. Southwest University, Chongqing, China, 2018.

[38]ZHANG L, HU B, LI W, Che R, et al. *OsPT2*, a phosphate transporter, is involved in the active uptake of selenite in rice[J]. The New phytologist, 2014, 201(4):1183-1191.

[39]ZHANG Q L, ZHANG F X, FENG Z X, et al. The preliminary experimental study on introduction and cultivation of tree tomato [J]. Shandong Forestry Science and Technology, 2005, 35(3):25-26.